Lecture Notes in Artificial Intelligence 13062

Subseries of Lecture Notes in Computer Science

Series Editors

Randy Goebel
University of Alberta, Edmonton, Canada
Yuzuru Tanaka
Hokkaido University, Sapporo, Japan
Wolfgang Wahlster
DFKI and Saarland University, Saarbrücken, Germany

Founding Editor

Jörg Siekmann
DFKI and Saarland University, Saarbrücken, Germany

More information about this subseries at http://www.springer.com/series/1244

Luis Espinosa-Anke · Carlos Martín-Vide ·
Irena Spasić (Eds.)

Statistical Language and Speech Processing

9th International Conference, SLSP 2021
Cardiff, UK, November 23–25, 2021
Proceedings

Springer

Editors
Luis Espinosa-Anke ⓘD
Cardiff University
Cardiff, UK

Carlos Martín-Vide ⓘD
Rovira i Virgili University
Tarragona, Spain

Irena Spasić ⓘD
School of Computer Science and Informatics
Cardiff University
Cardiff, UK

ISSN 0302-9743   ISSN 1611-3349   (electronic)
Lecture Notes in Artificial Intelligence
ISBN 978-3-030-89578-5   ISBN 978-3-030-89579-2   (eBook)
https://doi.org/10.1007/978-3-030-89579-2

LNCS Sublibrary: SL7 – Artificial Intelligence

This Springer imprint is published by the registered company Springer Nature Switzerland AG
The registered company address is: Gewerbestrasse 11, 6330 Cham, Switzerland

Preface

These proceedings contain the papers that were presented at the 9th International Conference on Statistical Language and Speech Processing (SLSP 2021), held in Cardiff, UK, during November 23–25, 2021, in conjunction with SLSP 2020, which had to be postponed due to the COVID-19 pandemic.

The scope of SLSP deals with topics of either theoretical or applied interest discussing the employment of statistical models (including machine learning) within language and speech processing, namely the following:

- Anaphora and coreference resolution
- Audio event detection
- Authorship identification, plagiarism, and spam filtering
- Biases, explainability, and interpretability in language and speech processing
- Corpora and resources for speech and language
- Data mining, term extraction, and semantic web
- Dialogue systems and spoken language understanding
- Information retrieval and information extraction
- Knowledge representation and ontologies
- Lexicons and dictionaries
- Machine translation and computer-aided translation
- Multimodal technologies
- Natural language understanding and generation
- Neural representation of speech and language
- Opinion mining and sentiment analysis
- Part-of-speech tagging, parsing, and semantic role labeling
- Question-answering systems for speech and text
- Speaker identification and verification
- Speech recognition and synthesis
- Spelling correction
- Text categorization and summarization
- Text normalization and inverted text normalization
- Text-to-speech
- User modeling
- Wake word detection

SLSP 2021 received 21 submissions. Every paper was reviewed by three Program Committee members. There were also a few external experts consulted. After a thorough and vivid discussion phase, the committee decided to accept nine papers (which represents an acceptance rate of about 43%). The conference program included two invited talks and some poster presentations of work in progress as well.

The excellent facilities provided by the EasyChair conference management system allowed us to deal with the submissions properly and handle the preparation of these proceedings in time.

We would like to thank all invited speakers and authors for their contributions, the Program Committee and the external reviewers for their diligent cooperation, and Springer for its very professional publishing work.

September 2021   Luis Espinosa-Anke
                                                                   Carlos Martín-Vide
                                                                       Irena Spasić

Organization

Organizing Committee

Luis Espinosa-Anke	Cardiff University, UK
Sara Morales	IRDTA Brussels, Belgium
Manuel Parra-Royón	University of Granada, Spain
David Silva	IRDTA London, UK
Irena Spasić	Cardiff University, UK

Program Committee

Mahmoud Al-Ayyoub	Jordan University of Science and Technology, Jordan
Chitta Baral	Arizona State University, USA
Jon Barker	University of Sheffield, UK
Jean-François Bonastre	University of Avignon, France
Fethi Bougares	University of Le Mans, France
Felix Burkhardt	audEERING, Germany
Nicoletta Calzolari	National Research Council, Italy
Bill Campbell	Amazon, USA
Angel Chang	Simon Fraser University, Canada
Kenneth W. Church	Baidu Research, USA
Marcello Federico	Amazon AI, USA
Robert Gaizauskas	University of Sheffield, UK
Ondřej Glembek	Brno University of Technology, Czech Republic
Thomas Hain	University of Sheffield, UK
Gareth Jones	Dublin City University, Ireland
Martin Karafiát	Brno University of Technology, Czech Republic
Philipp Koehn	Johns Hopkins University, USA
Carlos Martín-Vide	Rovira i Virgili University, Spain (Chair)
Seiichi Nakagawa	Chubu University, Japan
Stephen Pulman	University of Oxford, UK
Matthew Purver	Queen Mary University of London, UK
Paolo Rosso	Technical University of Valencia, Spain
Diana Santos	University of Oslo, Norway
Irena Spasić	Cardiff University, UK
Tomek Strzalkowski	Rensselaer Polytechnic Institute, USA
Tomoki Toda	Nagoya University, Japan
Isabel Trancoso	Instituto Superior Técnico, Portugal
K. Vijay-Shanker	University of Delaware, USA
Hsin-Min Wang	Academia Sinica, Taiwan
Andy Way	Dublin City University, Ireland
Wlodek Zadrozny	University of North Carolina Charlotte, USA

Additional Reviewers

Lorenc, Petr
Szoke, Igor

Contents

Language

Improving German Image Captions Using Machine Translation and Transfer Learning

Rajarshi Biswas[1]([✉]), Michael Barz[1,2], Mareike Hartmann[1],
and Daniel Sonntag[1,2]

[1] German Research Center for Artificial Intelligence (DFKI),
Saarland Informatics Campus D3_2, 66123 Saarbruecken, Germany
{rajarshi.biswas,michael.barz,mareike.hartmann,daniel.sonntag}@dfki.de
[2] Applied Artificial Intelligence, Oldenburg University, Marie-Curie Str. 1,
26129 Oldenburg, Germany

Abstract. Image captioning is a complex artificial intelligence task that involves many fundamental questions of data representation, learning, and natural language processing. In addition, most of the work in this domain addresses the English language because of the high availability of annotated training data compared to other languages. Therefore, we investigate methods for image captioning in German that transfer knowledge from English training data. We explore four different methods for generating image captions in German, two baseline methods and two more advanced ones based on transfer learning. The baseline methods are based on a state-of-the-art model which we train using a translated version of the English MS COCO dataset and the smaller German Multi30K dataset, respectively. Both advanced methods are pre-trained using the translated MS COCO dataset and fine-tuned for German on the Multi30K dataset. One of these methods uses an alternative attention mechanism from the literature that showed a good performance in English image captioning. We compare the performance of all methods for the Multi30K test set in German using common automatic evaluation metrics. We show that our advanced method with the alternative attention mechanism presents a new baseline for German BLEU, ROUGE, CIDEr, and SPICE scores, and achieves a relative improvement of 21.2% in BLEU-4 score compared to the current state-of-the-art in German image captioning.

Keywords: Natural language understanding and generation · Multimodal technologies · Image captioning · Natural language processing

Image captioning, i.e., the task of automatically describing an image, is an interesting problem of artificial intelligence research. It is multimodal in nature and lies at the intersection of computer vision and natural language processing. The

© Springer Nature Switzerland AG 2021
L. Espinosa-Anke et al. (Eds.): SLSP 2021, LNAI 13062, pp. 3–14, 2021.
https://doi.org/10.1007/978-3-030-89579-2_1

problem has witnessed rapid progress in the last few years owing to the development of novel deep neural architectures, training procedures, rapid advance-
ment in GPU computing power and lastly the availability of large annotated
datasets. However, the vast majority of research in this domain concentrates on
the English language. The primary reason for this development is the high availability of annotated image captioning datasets in English compared to other languages. For instance, the English MS COCO dataset [19] contains 164,063 images
each with 5 accompanying captions totaling to 820,315 captions. In comparison,
the Multi30K [9] dataset, which includes German captions sourced from native
speakers, contains 31,014 images with 155,070 accompanying German captions.
This is almost one order less in size than the MS COCO dataset. This sparsity
of resources is a major obstacle in developing effective neural models for caption
generation in German or other non English languages. As a result, there is a gap
in research on image caption generation in German. It is studied mostly as a
sub-task of multimodal machine translation where the image provides additional
information for the translation task. Elliott et al. [9] introduced the first dedicated German image captioning dataset, Multi30K, sourced from native German
speakers. Jaffe [15] studied the problem of generating image descriptions in German. For this purpose, they explored different model architectures which use a
training corpus containing captions in both English and German. They generate
captions for both languages, but discard the English output. Their best approach, which is based on an attention pipeline with random embeddings, is the
current state-of-the-art in producing German image captions.

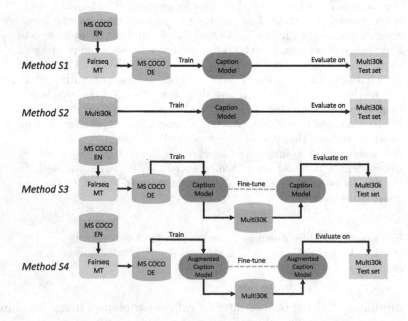

Fig. 1. The methods for German image caption generation that we compare in this
work.

In this work, we aim at improving caption generation in German by utilizing the large-scale MS COCO dataset in English. This is different from Jaffe [15] who used the Multi30K dataset for model training only. We transfer the English resources by translating all captions to German using the state-of-the-art neural machine translator Fairseq [21]. This way, we distantly leverage the higher availability of resources in the English-to-German translation domain. In total, we compare four different methods for generating German image captions on the Multi30K test set (see Fig. 1). We include two baseline methods and two more advanced methods based on fine-tuning. All methods are based on an adapted version of the encoder-decoder based neural architecture described in [3,5]. The baseline models are trained on the translated MS COCO dataset (*S1*) and the train set of the Multi30K dataset (*S2*), respectively. For both advanced methods, we pre-train the model on the translated MS COCO dataset and fine-tune it using the train set of the Multi30K dataset. We use the same model as for the baseline methods (*S3*) and a model with an alternative attention mechanism as described in [4] (*S4*). We hypothesize that both fine-tuning methods perform better than the baseline methods in terms of common evaluation metrics. Also, we expect that the alternative attention mechanism *S4* further improves the image caption quality and beats the current state-of-the-art by Jaffe [15].

The rest of the paper is structured as follows: We discuss the progress in multilingual caption generation in Sect. 1. Then we discuss the technical details of our approach in Sect. 2 followed by a detailed report of our evaluation and its results in Sect. 3. We discuss the results in Sect. 4 and conclude our paper in Sect. 5.

1   Related Work

Approaches for multilingual image captioning can be divided into two broad categories: translation-based approaches and alignment-based approaches. Translation-based approaches rely on machine translation models to either translate generated captions to the target language or to create an image captioning dataset in the target language for training language-specific models. Elliott et al. [7] are one of the first to study the task of multilingual image caption generation. They use features from both source and target language model and generate the captions using an LSTM [13] based decoder. Hitschler et al. [12] translate image captions from one language to the other using the image as additional input. This image guided translation is the focus of the WMT 2016 multimodal machine translation task [23] and the WMT 2017 task [8] with some variations, such as, unavailability of the source language at test time. These WMT tasks on multimodal machine translation find that purely text-based machine translation techniques provide a strong baseline when translating captions from one language to another. Additionally, they found that supplementing machine translation techniques with information from the image only results in a marginal improvement. For example, the work in Huang et al. [14] re-ranked the translation output using image features, but could not improve the METEOR score compared to their baseline. This trend is also observed for the task of generating cross-lingual image descriptions in WMT 2016. In spite of using attention based models, the image does not provide much benefit towards generating captions in German and all the highest scoring systems in WMT 2016 for the cross-lingual image description multimodal task ignored the image. In the WMT 2017 task this observation is repeated, that is, text-only systems perform better and obtain higher scores compared to multimodal systems that use images as context. Jaffe [15] generate image captions in German as part of the WMT 2017 multimodal translation sub task on multilingual image caption generation. They use the Multi30K dataset for this purpose and explore different neural architectures. They use the images with both English and German captions for training their models. In fact, they generate captions in both languages, but discard the English output during evaluation. Also, they experiment with textual attention for caption generation. Their architecture that uses attention over the German caption output achieves the highest scores in terms of the BLEU-4 and METEOR metrics.

Alignment-based approaches rely on a joint embedding space. These methods aim to first learn an alignment between the given image and corresponding English captions in a common latent space. This alignment is then used to relate to the target language. They assume better alignment leads to better captions generated in the target language. Through this process they try to make up for the lack of annotated training data in the target language. For instance, Miyazaki et al. [20] pre-train a captioning model on the MS COCO dataset. Later they modify this model and train the modified model on Japanese data for generating captions. Wu et al. [26] combine merits from both, alignment-based and translation-based approaches, for multilingual image captioning in a unified architecture. In their work, given an input image, they generate English captions and, then, the caption in the target language. Similarly, Thapliyal et al. [24] propose a system that uses existing English annotations and their translations at training time. At run time their system generates an English caption and then a corresponding caption in the target language. Lan et al. [17] propose a fluency guided framework where they aim to learn a cross-lingual captioning model from machine translated sentences. Their proposed framework automatically estimates the fluency of the sentences and uses the estimated fluency scores as part of the cost function to train an image captioning model for the target language. The work of Gu et al. [10] first uses a pivot language for capturing the characteristics of the image captioner and then uses a pivot-target language parallel corpus to align the image captioner to the target language.

Our advanced methods can be classified as translation-based, because we use the Fairseq neural machine translator to translate the MS COCO dataset into German. However, our approach differs from previous works on translation based methods in two key aspects. First, we use a fine-tuning process where we pre-train our captioning model on the translated dataset and subsequently fine tune the model on the German captioning dataset, Multi30K, sourced from native speakers of the language. We assume this process can help in learning language specific nuances through this process. And, it is much simpler compared to approaches using a pivot language. Second, we apply a modified attention scheme [4] that has been shown to improve caption generation for English.

2 Method

We implement four methods for generating image captions in German based on the neural image captioning model presented in [4,5]. We include two baseline models and two advanced models based on fine-tuning (see Fig. 1). The baseline models use the MS COCO dataset translated into German and the Multi30K dataset respectively for training the captioning model. For the advanced models, we pre-train the caption model using the translated MS COCO dataset and then fine-tune it on the Multi30K dataset. Also, the baseline methods *S1*, *S2* and the method *S3* use the caption model from [5,27]. In contrast, in the advanced method *S4* we fine-tune the image captioning model with the more effective augmented attention mechanism proposed in [4].

2.1 Image Captioning Datasets

We translate the original MS COCO dataset [19] from English into German using the Fairseq neural machine translator. The translated MS COCO dataset contains 82,783 images with 5 corresponding captions in the training set while the validation set contain 5,000 images each with 5 groundtruth captions per image. We refer to this data split as the *COCO_Split*. We also use the Multi30K German image captioning dataset for training the captioning models in our methods. For Multi30K, the training set contains 29,000, the validation set contains 1014 and the test set contains 1000 images respectively with 5 corresponding captions per image. We denote this break up as the *M30k_Split*.

2.2 Image Captioning Model

For all methods, we use the neural encoder-decoder model with visual attention mechanism adapted from [3,5,27]. The image encoder part of this model is based on the ResNet-101 model with 101 layers [11]. We do not perform any pre-processing on the images. We apply spatially adaptive max-pooling which results in a fixed size output of $14 \times 14 \times 2048$ for each image. Thus, each image is encoded as 196 vectors with a dimension of 2048. The decoder in our caption generation model is an LSTM [13], and we build our vocabulary by dropping word types with a frequency <5. We set the dimensions for the LSTM hidden state, image, word and attention embeddings to 512 and train the model under the cross entropy objective, using the ADAM [16] optimizer. All models are trained for 30 epochs, followed by 30 epochs of fine-tuning for methods *S3* and *S4*.

2.3   Caption Generation Methods

To build our baseline method $S1$, we train the caption generation model as described above using the translated MS COCO dataset. Our baseline method $S2$ is trained using the $M30k\_Split$. For $S3$, we pre-train the model on the translated German MS COCO dataset using $COCO\_Split$. This allows the model to learn the initial mapping from images to the German language from the translated corpus. Subsequently, we fine-tune this model on the Multi30K dataset, sourced from native speakers, using the $M30k\_Split$. For $S4$, we use the attention mechanism presented in [4], which has been shown to improve English captioning systems. This mechanism incorporates object-specific localized maps from a region proposal network for this purpose. Specifically, we represent an input image I as a set of feature vectors, $I = \{f_1, f_2, ..., f_n\}$ where $f_i \in \mathbb{R}^d$. Each element in this set represents the encoding of a bounding box detected by a region proposal network that is encoded using the ResNet-101 model [11]. We extract the image regions inside the final bounding boxes obtained after non-maxima suppression and embed them into the feature space learned by ResNet-101 pre-trained on the ImageNet [6] dataset. We set a high threshold (0.8) for the classification probability for the regions to be selected. Subsequently, we compute visual attention on the joint embedding space formed by the union of high-level features obtained from the encoder of the caption generator and the low-level features obtained from the object specific local regions of the input image. We use 10 additional feature vectors for every image to represent the local regions. So, our attention mechanism at every time-step produces a mask over 206 spatial locations. This mask is applied to a set of image features and the result is spatially averaged to produce a 2048 dimensional representation of the attended portion of the image. We pre-train this caption model with the augmented attention mechanism first on the translated MS COCO dataset using $COCO\_Split$ and then fine-tune the trained model on the Multi30K dataset with $M30k\_Split$ for learning language specific nuances. For a quick reference all considered methods are listed below.

1. ($S1$) we train the caption generation model using only the translated MS COCO dataset.
2. ($S2$) we train the caption generation model using only the Multi30K dataset.
3. ($S3$) we train the image caption generation model on the translated MS COCO dataset and then fine tune the model on the Multi30K dataset.
4. ($S4$) we train the image caption model with augmented attention on the translated MS COCO dataset and then fine tune the model on the Multi30K dataset.

3   Evaluation

We test all methods, explained above, using the Multi30K test set and compare the generated captions using automated metrics commonly used in the image captioning research community (see Sect. 3.1). Our goal is to ascertain the most effective method for generating image captions in German. In this regard, we investigate the effect of pre-training a model on the translated MS COCO dataset and the impact of using the alternative attention mechanism on the quality of generated image captions. Also, we compare the scores of all four methods to the results reported in [15] as they achieved the highest metric scores for caption generation in German.

3.1 Metrics

We compute a group of automated metrics commonly used in the image captioning research community: BLEU [22], METEOR [2], ROUGE [18], CIDEr [25] and SPICE [1]. These metrics primarily focus on the n-gram overlap between the generated and ground-truth captions. For convenience, we provide a short description for each metric. BLEU scores are computed by directly matching n-grams between individual machine generations and a corresponding set of ground-truth references. It is always between 0 and 1 where 0 indicates no overlap and 1 indicates a perfect overlap. Depending on the size of the n-grams you get different BLEU scores, i.e., BLEU-1, BLEU-2, BLEU-3, BLEU-4. METEOR evaluates outputs from a machine translation system. It computes the harmonic mean of unigram precision and recall. Recall is weighted higher than precision. ROUGE measures the longest matching sequence of words. An advantage of it is that it does not require consecutive matches but in-sequence matches that reflects sentence level order. CIDEr measures the similarity of a generated sentence against a set of ground truth sentences composed by humans and shows high agreement with consensus as assessed by humans. SPICE denotes semantic propositional image caption evaluation. It uses semantic information in the form of a scene graph to measure the similarity between the ground-truth and machine generated captions.

3.2 Hypothesis

We hypothesize that the advanced methods, *S3* and *S4*, yield a better performance than the baseline methods *S1*, *S2* that do not use fine-tuning in terms of the metrics mentioned above. We expect that the method (*S4*) generates the best German captions compared to our other methods, but also to the current state-of-the-art performance in German image captioning as reported in [15] which provides baseline scores for BLEU-4 and METEOR.

3.3 Results

The scores for all German image caption generation methods are summarized in Table 1. The scores are computed using the standard metric computation package which ensures comparability with Jaffe [15]. Among our methods, *S4* yields the best scores: it achieves higher scores for BLEU-1,2,3,4, ROUGE, CIDEr, and SPICE metrics. Only the METEOR score obtained with *S4* is lower by a small

(a) **S4:** ein mann mit hut und sonnen-brille sitzt auf einem felsen und schaut auf das wasser (*a man in a hat and sunglasses is sitting on a rock and looking at the water*); **S3:** ein mann mit mütze sitzt auf einem felsen und schaut auf sein handy (*a man in a hat is sitting on a rock and looking at his mobile phone*); **S2:** ein mann mit hut sitzt auf einem skateboard (*a man in a hat is sitting on a skateboard*); **S1:** ein mann sitzt auf einem UNK (*a man is sitting on a UNK*)

(b) **S4:** eine gruppe von menschen sitzt an einem tisch mit essen (*a group of people is sitting at a table with food*); **S3:** eine frau und ein mann sitzen an einem tisch und essen kuchen (*a woman and a man are sitting at a table and eating cake*); **S2:** eine gruppe von menschen sitzt an einem tisch mit einem tisch (*a group of people is sitting at a table with a table*); **S1:** zwei frauen sitzen an einem tisch und essen (*two women are sitting at a table and eating*)

(c) **S4:** ein mann sitzt an einem tisch und schreibt etwas auf ein papier (*a man is sitting at a table and writing something on a paper*); **S3:** ein mann sitzt an einem tisch und schreibt in ein heft (*a man is sitting at a table and writing something in a notebook*); **S2:** ein mann sitzt an einem tisch mit einem laptop (*a man is sitting at a table with a laptop*); **S1:** zwei männer sitzen an einem tisch und spielen (*two men are sitting at a table and playing a game*)

(d) **S4:** ein mann klettert an einem seil gesichert eine felswand hinauf (*a man is climbing up a rock face secured by a rope*); **S3:** ein mann klettert an einem seil gesichert an einem seil (*a man is climbing on a rope secured by a rope*); **S2:** ein mann klettert an einem felsen (*a man is climbing up a rock*); **S1:** ein mann fährt auf einem UNK durch eine UNK (*a man is driving on a UNK through a UNK*)

Fig. 2. Example German image captions generated with the methods explored in our work. Italics in brackets provide English translations of the generated captions.

margin of 0.006 compared to *S3* and by 0.003 than *S2*. We observe that all metric scores, apart from METEOR, gradually increase from *S1* to *S4*. This trend also extends to the method (*S3*) and the method (*S4*). Also, the BLEU-4 score of *S4* is better than the corresponding score reported in the current state-of-the-art approach by Jaffe [15] by an absolute margin of 0.025 that is a relative improvement of 21.19%. However, our METEOR score is lower by 0.048. We use the same technique as Jaffe to compute the metrics and believe this inconsistency could be due to the low correlation between BLEU and METEOR as observed by Jaffe [15]. Unfortunately, the authors did not report other metrics for image captioning like CIDEr and SPICE for which our approach *S4* obtains highest scores among our methods.

Table 1. Performance scores of different methods used for generating German image captions on the Multi30K test set.

Strategy	Bleu-1	Bleu-2	Bleu-3	Bleu-4	METEOR	ROUGE	CIDEr	SPICE
Jaffe [15]	–	–	–	0.118	**0.205**	–	–	–
S4	**0.527**	**0.352**	**0.227**	**0.143**	0.157	**0.369**	**0.307**	**0.035**
S3	0.508	0.317	0.191	0.107	0.163	0.358	0.250	0.029
S2	0.482	0.297	0.178	0.101	0.160	0.351	0.227	0.027
S1	0.456	0.270	0.151	0.081	0.151	0.326	0.177	0.023

4 Discussion

The results obtained in our work (see Table 1) show that the method (*S4*) with the alternative attention mechanism results in higher BLEU-4 score compared to the value reported in the state-of-the-art approach [15] for German image captioning, indicating that our hypothesis could be confirmed in terms of the BLEU-4 metric. The comparison of *S4* with *S3, S2, S1* establishes the merit in using the augmented attention mechanism. This is also observed in the examples shown in Fig. 2 which shows that the captions generated using *S4* are comparatively better than the other methods. Also, the captions generated using *S3, S2, S1* do not capture the relevant details in the image compared to *S4*. Our results also show the benefit of pre-training the caption generation model on the translated MS COCO dataset followed by fine-tuning it on the smaller Multi30K dataset. Importantly, there is a consistent gradual increase in the BLEU, ROUGE, CIDEr, and SPICE scores as we transition from *S1* to *S4*. A comparison of the scores from *S1* and *S2* shows that the Multi30K training data, sourced from native German speakers, is more influential to the caption generation model compared to the machine translated German MS COCO dataset in our test setting. Using *S3*, we show that pre-training followed by fine tuning could be one of the possible ways to overcome the requirement of large amount of annotated data for training an image captioning model in German as it achieves better scores compared to both *S1* and *S2* across all the metrics. Finally, we show that training the caption generation model with the augmented attention mechanism using fine-tuning in *S4* results in highest improvement relative to all the strategies we used in our work. This is evidenced through higher BLEU-1, 2, 3, 4, ROUGE, CIDEr and SPICE scores compared to those obtained by *S1*, *S2*, *S3*. Moreover, *S4* even obtains higher BLEU-4 score compared to the current state-of-the-art in German image captioning.

5   Conclusion

In this work, we implemented and evaluated four methods for caption generation in German with the goal of achieving state-of-the-art performance. We showed that our methods could serve as possible ways of overcoming the problem of sparse availability of training data for image captioning in the German language. Our best performing method uses an alternative attention mechanism from the literature [4] and leverages the vast resources available in English, i.e., the MS COCO dataset for cross-lingual information transfer in the context of image captioning via the Fairseq neural machine translator. The model is pre-trained on the translated MS COCO dataset and fine-tuned on the German Multi30K dataset sourced from native speakers. This model achieves the best BLEU-1, 2, 3, 4, ROUGE, CIDEr, and SPICE scores compared to our baseline methods. Moreover, the model with alternative attention mechanism obtained a higher BLEU-4 score than the state-of-the-art approach by Jaffe [15] by an absolute margin of 0.025 that is a relative improvement of 21.19%.

Acknowledgments. This work was funded by the German Federal Ministry of Research (BMBF) under grant number 01IW20005 (XAINES).

References

1. Anderson, P., Fernando, B., Johnson, M., Gould, S.: SPICE: semantic propositional image caption evaluation. In: Leibe, B., Matas, J., Sebe, N., Welling, M. (eds.) ECCV 2016. LNCS, vol. 9909, pp. 382–398. Springer, Cham (2016). https://doi.org/10.1007/978-3-319-46454-1_24
2. Banerjee, S., Lavie, A.: METEOR: an automatic metric for MT evaluation with improved correlation with human judgments. In: Proceedings of the ACL Workshop on Intrinsic and Extrinsic Evaluation Measures for Machine Translation and/or Summarization, pp. 65–72 (2005)
3. Biswas, R.: Diverse image caption generation and automated human judgement through active learning. Master's thesis, Saarland University (2019)
4. Biswas, R., Barz, M., Sonntag, D.: Towards explanatory interactive image captioning using top-down and bottom-up features, beam search and re-ranking. KI - Künstliche Intell. German J. Artif. Intell. - Organ Fachbereiches "Künstliche Intell." Gesellschaft für Inf. e.V. (KI) **34**(4), 571–584 (2020). https://doi.org/10.1007/s13218-020-00679-2

5. Biswas, R., Mogadala, A., Barz, M., Sonntag, D., Klakow, D.: Automatic judgement of neural network-generated image captions. In: Martín-Vide, C., Purver, M., Pollak, S. (eds.) SLSP 2019. LNCS (LNAI), vol. 11816, pp. 261–272. Springer, Cham (2019). https://doi.org/10.1007/978-3-030-31372-2_22
6. Deng, J., Dong, W., Socher, R., Li, L.J., Li, K., Fei-Fei, L.: ImageNet: a large-scale hierarchical image database. In: 2009 IEEE Conference on Computer Vision and Pattern Recognition, pp. 248–255. IEEE (2009)
7. Elliott, D., Frank, S., Hasler, E.: Multilingual image description with neural sequence models. arXiv Computation and Language (2015)
8. Elliott, D., Frank, S., Barrault, L., Bougares, F., Specia, L.: Findings of the second shared task on multimodal machine translation and multilingual image description. In: Proceedings of the Second Conference on Machine Translation, pp. 215–233. Association for Computational Linguistics, Copenhagen, Denmark, September 2017. https://doi.org/10.18653/v1/W17-4718. https://www.aclweb.org/anthology/W17-4718
9. Elliott, D., Frank, S., Sima'an, K., Specia, L.: Multi30K: multilingual English-German image descriptions. In: Proceedings of the 5th Workshop on Vision and Language, pp. 70–74. Association for Computational Linguistics, Berlin, August 2016. https://doi.org/10.18653/v1/W16-3210. https://www.aclweb.org/anthology/W16-3210
10. Gu, J., Joty, S., Cai, J., Wang, G.: Unpaired image captioning by language pivoting. In: Ferrari, V., Hebert, M., Sminchisescu, C., Weiss, Y. (eds.) ECCV 2018, Part I. LNCS, vol. 11205, pp. 519–535. Springer, Cham (2018). https://doi.org/10.1007/978-3-030-01246-5_31
11. He, K., Zhang, X., Ren, S., Sun, J.: Deep residual learning for image recognition. In: 2016 IEEE Conference on Computer Vision and Pattern Recognition (CVPR), pp. 770–778 (2016). https://doi.org/10.1109/CVPR.2016.90
12. Hitschler, J., Schamoni, S., Riezler, S.: Multimodal pivots for image caption translation. In: Proceedings of the 54th Annual Meeting of the Association for Computational Linguistics (Volume 1: Long Papers), pp. 2399–2409. Association for Computational Linguistics, Berlin, August 2016. https://doi.org/10.18653/v1/P16-1227. https://www.aclweb.org/anthology/P16-1227
13. Hochreiter, S., Schmidhuber, J.: Long short-term memory. Neural Comput. **9**(8), 1735–1780 (1997). https://doi.org/10.1162/neco.1997.9.8.1735
14. Huang, P.Y., Liu, F., Shiang, S.R., Oh, J., Dyer, C.: Attention-based multimodal neural machine translation. In: Proceedings of the First Conference on Machine Translation: Volume 2, Shared Task Papers, pp. 639–645. Association for Computational Linguistics, Berlin, August 2016. https://doi.org/10.18653/v1/W16-2360. https://www.aclweb.org/anthology/W16-2360
15. Jaffe, A.: Generating image descriptions using multilingual data. In: Proceedings of the Second Conference on Machine Translation, pp. 458–464. Association for Computational Linguistics, Copenhagen, September 2017. https://doi.org/10.18653/v1/W17-4750. https://www.aclweb.org/anthology/W17-4750
16. Kingma, D.P., Ba, J.: Adam: a method for stochastic optimization. In: 3rd International Conference on Learning Representations, ICLR 2015, San Diego, CA, USA, 7–9 May 2015. Conference Track Proceedings (2015). http://arxiv.org/abs/1412.6980
17. Lan, W., Li, X., Dong, J.: Fluency-guided cross-lingual image captioning. In: Proceedings of the 25th ACM International Conference on Multimedia, MM 2017, pp. 1549–1557. Association for Computing Machinery, New York (2017). https://doi.org/10.1145/3123266.3123366

18. Lin, C.: ROUGE: a package for automatic evaluation of summaries. Text Summarization Branches Out (2004)
19. Lin, T.-Y., et al.: Microsoft COCO: common objects in context. In: Fleet, D., Pajdla, T., Schiele, B., Tuytelaars, T. (eds.) ECCV 2014. LNCS, vol. 8693, pp. 740–755. Springer, Cham (2014). https://doi.org/10.1007/978-3-319-10602-1_48
20. Miyazaki, T., Shimizu, N.: Cross-lingual image caption generation. In: Proceedings of the 54th Annual Meeting of the Association for Computational Linguistics (Volume 1: Long Papers), pp. 1780–1790. Association for Computational Linguistics, Berlin, August 2016. https://doi.org/10.18653/v1/P16-1168. https://www.aclweb.org/anthology/P16-1168
21. Ott, M., et al.: fairseq: a fast, extensible toolkit for sequence modeling. In: Proceedings of the 2019 Conference of the North American Chapter of the Association for Computational Linguistics (Demonstrations), pp. 48–53. Association for Computational Linguistics, Minneapolis, June 2019. https://doi.org/10.18653/v1/N19-4009. https://www.aclweb.org/anthology/N19-4009
22. Papineni, K., Roukos, S., Ward, T., Zhu, W.: BLEU: a method for automatic evaluation of machine translation, pp. 311–318. Association for Computational Linguistics (2002)
23. Specia, L., Frank, S., Sima'an, K., Elliott, D.: A shared task on multimodal machine translation and crosslingual image description. In: Proceedings of the First Conference on Machine Translation: Volume 2, Shared Task Papers, pp. 543–553. Association for Computational Linguistics, Berlin, August 2016. https://doi.org/10.18653/v1/W16-2346. https://www.aclweb.org/anthology/W16-2346
24. Thapliyal, A.V., Soricut, R.: Cross-modal language generation using pivot stabilization for web-scale language coverage. In: Proceedings of the 58th Annual Meeting of the Association for Computational Linguistics, pp. 160–170. Association for Computational Linguistics, Online, July 2020. https://doi.org/10.18653/v1/2020.acl-main.16. https://www.aclweb.org/anthology/2020.acl-main.16
25. Vedantam, R., Zitnick, C., Parikh, D.: CIDEr: consensus-based image description evaluation. In: Computer Vision and Pattern Recognition, pp. 4566–4575 (2015)
26. Wu, Y., Zhao, S., Chen, J., Zhang, Y., Yuan, X., Su, Z.: Improving captioning for low-resource languages by cycle consistency. In: 2019 IEEE International Conference on Multimedia and Expo (ICME), pp. 362–367 (2019). https://doi.org/10.1109/ICME.2019.00070
27. Xu, K., et al.: Show, attend and tell: neural image caption generation with visual attention. In: Bach, F., Blei, D. (eds.) Proceedings of the 32nd International Conference on Machine Learning. Proceedings of Machine Learning Research, vol. 37, pp. 2048–2057. PMLR, Lille, 07–09 July 2015. http://proceedings.mlr.press/v37/xuc15.html

Automatic News Article Generation from Legislative Proceedings: A Phenom-Based Approach

Anastasiia Klimashevskaia[1] , Richa Gadgil[2] , Thomas Gerrity[2] ,
Foaad Khosmood[2(✉)] , Christian Gütl[1] , and Patrick Howe[2]

[1] Graz University of Technology, Graz, Austria
a.klimashevskaia@student.tugraz.at, c.guetl@tugraz.at
[2] California Polytechnic State University, San Luis Obispo, USA
{rgadgil,tmgerrit,foaad,pchowe}@calpoly.edu

Abstract. Algorithmic journalism refers to automatic AI-constructed news stories. There have been successful commercial implementations for news stories in sports, weather, financial reporting and similar domains with highly structured, well defined tabular data sources. Other domains such as local reporting have not seen adoption of algorithmic journalism, and thus no automated reporting systems are available in these categories which can have important implications for the industry. In this paper, we demonstrate a novel approach for producing news stories on government legislative activity, an area that has not widely adopted algorithmic journalism. Our data source is state legislative proceedings, primarily the transcribed speeches and dialogue from floor sessions and committee hearings in US State legislatures. Specifically, we create a library of potential events called phenoms. We systematically analyze the transcripts for the presence of phenoms using a custom partial order planner. Each phenom, if present, contributes some natural language text to the generated article: either stating facts, quoting individuals or summarizing some aspect of the discussion. We evaluate two randomly chosen articles with a user study on Amazon Mechanical Turk with mostly Likert scale questions. Our results indicate a high degree of achievement for accuracy of facts and readability of final content with 13 of 22 users in the first article and 19 of 20 subjects of the second article agreeing or strongly agreeing that the articles included the most important facts of the hearings. Other results strengthen this finding in terms of accuracy, focus and writing quality.

Keywords: Algorithmic journalism · Natural language generation · Automatic summarization · Partial order planning · Artificial intelligence · Digital government

1 Introduction and Motivation

We present a novel method to generate news stories on state legislative proceedings. Our strategy is to generate algorithmic news content that summarize the

© Springer Nature Switzerland AG 2021
L. Espinosa-Anke et al. (Eds.): SLSP 2021, LNAI 13062, pp. 15–26, 2021.
https://doi.org/10.1007/978-3-030-89579-2_2

important events of a single discussion based on records of meetings and other data sources. Such content could be distributed to news organizations which could in turn print them in their publications or use them as a basis for writing more detailed stories.

The primary input data into this system are high quality, human-verified transcripts from hearings in state legislatures as developed by the Digital Democracy project [2] and already used in several other works [3,12,26,27]. Hearings are divided into "bill discussions", or dialogues about a specific legislation that are on average about 20 min of multi-speaker discussion, typically followed by a vote. In this paper we use a portion of the data consisting of transcripts from 5198 bill discussions containing 40788 individual speeches covering the California Legislature 2015–2017.

Our system generates articles based on abstractive summarizations of the bill discussions achieved using a planning system and a "phenom" (short for phenomenon) library that we also develop. We define and model the phenom as a software object with several standard attributes and methods. Each phenom is called by the planner to test for a certain specific event, condition or occurrence in the text of the discussion. If the search is successful, the phenom then contributes one or more sentences to the text of the article. The surface text realization is achieved through insertion of facts into pre-written English language sentence templates appropriately chosen for each phenom.

Due to strict requirements for high accuracy and traceability of all facts to original primary sources, we do not rely on predictive or transformer NLG models which are prone to "hallucinations" or tendency to generate inaccurate or nonsensical statements [10]. For quality purposes, we also verify the approach using human evaluations.

1.1 Motivation

News media play a crucial role in the functioning of democracies through coverage of governmental processes, but such coverage is in decline. The Pew Research Center found that the number of full-time reporters covering state legislatures fell by 35% between 2003 and 2014 [8]. Observers say the decline has likely accelerated in the years since the study was conducted [30]. The decline largely came from the ranks of newspapers, which have historically provided the greatest proportion of full-time statehouse reporters (only 15% of television stations send any reporters to cover state legislatures) [8]. Facing declining circulation and advertising revenues, U.S. newspapers shed 51% of their editorial employees between 2008 and 2019 [11].

With fewer staff, news outlets have decreased their coverage of governmental proceedings [1,6,18]. The decline in dedicated statehouse reporters has led to less coverage of policy and the state legislative process, with most remaining coverage focusing narrowly on legislative outcomes such as the final passage of a bill [33,34]. State legislatures themselves have responded by bulking up their own in-house media offerings via websites, video reports and social media [33].

1.2 Organization

The rest of this paper is organized as follows: in Sect. 2 we provide a short overview on the previous research and related works. Section 3 introduces first the concept, then details further the developments and technical issues of the project. It is followed by Sect. 4 with a description of the design and procedure of the user study. Section 4.2 is a discussion about our findings. Finally, Sect. 5 concludes the paper, giving some final ideas for future work.

2 Related Work

Summarization of dialog or meeting transcripts is a tricky task for programmers. Several experiments described applying classic extractive summarization methods on speech transcripts. It was concluded that the spontaneous nature of speech results in lower quality summaries compared to written text [5,10]. Meeting transcripts consist of unstructured utterances with long-range semantic dependencies [32]. They can contain more grammatical and spelling errors and are more noisy, thus producing a less readable and concise summary using extractive techniques [16,19]. Still, there are attempts to utilize an extractive approach [4,10,25,35] with transcripts. Template-based NLG was previously successfully utilized in news generation, with human written templates [20,29] as well as generated ones [23].

The approach of a flat facts set described by [13] with a certain addition to fact types and template adjustment served as a big inspiration for this project. This methodology in combination with hand-written templates as in [20,29] could provide good results in accuracy and grammatically of the final text. Event, entity and relation recognition is one of the good examples for fact extraction approaches. Works like [7,14,21] give great descriptions of such event detection systems, however, they are mostly concentrating on extraction from coherent written text, while our task is basing on transcripts of spoken language, which quite often lacks proper grammar, coherence and contains transcription errors, incomplete sentences, interruptions, etc. Nevertheless, these works still provide invaluable insights and ideas on event-entity-relation detection and extraction.

Our contribution is in the system of phenoms that combine event detection, template based realization for news articles and a modified partial-order planner through which a semantic structure of a news article is enforced.

3 Approach and Development

We design a system to ingest records of legislative bill discussions and produce an article as the main deliverable (see Fig. 1). Initially, all discussions are identified and extracted from the database by a newsworthiness filter, developed outside the focus of this paper to simply prioritize certain bill discussions over others. Transcripts then undergo preprocessing where every utterance from the transcript is tokenized, tagged for speaker, function, and named entities resulting in Fig. 2.

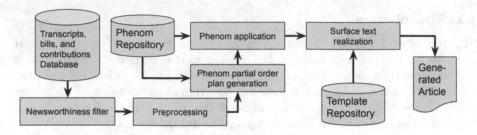

Fig. 1. The data flow diagram of the system

utterance_order	pid	text	Paragraph label	Secondary label
1	92	Everybody turn off their cell phones if they want to bother us while we spend the next 3 to 4 minutes in this committee. And lets take our roll. We have 4 measures, all on consent items. And all the items on our agenda are on consent. So we'll take the roll and then we have a motion on the consent calendar.	1	
2	2998	Senators Beall? Here. Beall present, Cannella? Here. Cannella present. Allen? Bates? Gaines? Galgiani? Leyva? Here. Leyva present. McGuire? Here. McGuire present. Mendoza? Here. Mendoza present. Roth? Here. Roth, present. Wieckowski? Here. Wieckowski present.	0	
3	92	Okay these consent calendar items are ACR 58 by various Assembly Members, Williams, ACR 63, Maienschein, and ACR 65, Brough, ACR 78, Salas, and if there's no discussion on these items, we'll have a motion for the approval of the consent calendar. The motion's to approve this consent calendar, and we'll take a vote.	2	4
4	2998	On the consent calendar, Senators Beall? Aye. Beall aye, Cannella? Aye. Cannella aye. Allen? Bates? Gaines? Galgiani? Leyva? Aye. Leyva aye. McGuire? Aye. McGuire aye. Mendoza? Aye. Mendoza aye. Roth? Aye. Roth aye. Wieckowski? Aye. Wieckowski aye.	4	
5	92	7? We're gonna put that on call, we have 7 votes in favor. So we'll wait for other members to come and record their votes, and as soon as they come and vote, we will adjourn the committee. So we're waiting on 4 Senators, and we'll wait for them and record their votes and adjourn.	0	
6	2998	Thank you Senator Beall.	0	
7	92	Thank you.	4	
8	92	Consent calendar.	2	
9	2998	Senators Allen? Bates? Gaines? Galgiani? Galgiani aye.	4	

Fig. 2. Example of a tagged transcript portion in preprocessing

3.1 Phenom System

Legislative proceedings vary greatly, but we find some useful patterns and data in hearing videos and their written transcripts. This inspired the adoption of the phenom approach - systematic extraction of the key phenomena from the transcript text and storing them in a collection of fact data types to used for generating language for the article.

We create a library of phenoms for content selection. Each phenom is an object with data and methods designed to look for a specific event in the bill discussion, and take some action if the event is found. For example does a greeting occur in this discussion? A phenom called "greetings" could look for specific language that constitutes a greeting according to hand-crafted criteria. If found, the phenom generates facts leading to a statement like: "Senator Monning greeted the committee". That sentence becomes part of the news article. Given many phenoms each dedicated to typical occurrences in legislative hearings, we can generate many observations.

The order of executions of the phenoms are not set ahead of time. Instead every phenom has a set of preconditions and postconditions. A modified partial order planner selects a phenom from the library based on preconditions and executes it. After execution, that phenom may contribute some more conditions to the current state of the article that may enable other phenoms to be activated.

In this way, some phenom output are guaranteed to be in specific places in the article, such as the beginning section or the ending section. They can also form chains of dependencies allowing much more conditioned observations when called for. For example a news story may have a short statement about the background of a certain expert witness generated by phenom #2, but only if the witness is first called to testify, which would have been observed by phenom #1 (the precondition of phenom #2).

Phenom ideas are created based on several sources: Fist, a study of 300 real published news articles that contained the name of a bill that was before the California Legislature at the time of the article publication. These provided some standard observations that journalists include in news articles about state legislative action. Second, some interesting data available to us from the database were made into phenoms. For example, we can automatically flag the first-ever bill authored by a certain legislator, or the first bill of the session by that legislator. This small point may be very important to a hometown news organization but not immediately apparent to a reporter. Lastly, some purely functional phenoms are created to help with content arrangement, to ensure, for example, that a pull-quote not appear until the middle section of an article. For a selection of some of the phenoms and their respective frequencies of occurrence see Table 1.

3.2   Illustrative Phenom Examples

While certain phenoms have mostly formal purpose and use the metadata provided by the database, the others are designed to search for complex patterns and work directly with the transcript text, utilizing classification or scoring algorithms. Metadata-based and functional phenoms occur in the texts more often due to higher likelihood of availability of the data, while the more refined ones tend to be rare and are triggered not as often (see Table 1). The following paragraphs give examples of such high-order phenoms present in the system.

Pull-Quote Extraction. One of the phenoms implemented in the pipeline is aimed at pull-quote extraction from the transcript text. A pull-quote is a key sentence, or phrase that is highlighted, shown with a bigger font, and a line of attribution at a central location in the printed article layout. Pull-quotes are a traditional part of news articles. The pull-quote phenom has a multi-step process in which a newsworthy quote is determined by filtering, labelling, and ranking all the discussion utterances from the transcript. Preprocessing involves removing utterances from subjects of non-interest (such as those spoken by committee heads, chairs and staff). Following preprocessing, we correlate each utterance with a dialogue label based on the Switchboard Corpus [22,31]. The Switchboard corpus is a collection of about 2,400 two-sided telephone dialogues among 543 speakers, with each utterance annotated using the Discourse Annotation and Markup System of Labeling (DAMSL) tag set.

We only consider utterances that have the following labels: "Statement-opinion", "Rhetorical-Questions", "Hedge", "Action-directive" and "Apology

Table 1. List of select phenoms and percentage triggered. A set of 50 random bill discussions was used for this purpose.

Phenom	Description	%
bill_name	Metadata: bill name query	100
subjects	Metadata: bill subject query	100
presenter	Detects the person presenting the bill	94
motion	Metadata: bill motion query	98
vote_result	Metadata: voting result query	98
attendance	Metadata: attendance query	22
intro_length	Detects intro part longer than certain threshold	40
chair	Detects the chair of the meeting	50
expert_testimony	Detects an expert testimony longer than a certain threshold	50
testimony_alignment	Adds alignment information from the metadata to the testimony info	34
bill_mentions	Detects mentions of any other bill names	56
back_and_forth	Detects a back-and-forth discussion	10
alignment	Metadata: general alignment of the people present	10
first_bill	Metadata: is it the first bill by the legislator	2
sigle_party_split	Metadata: split voting of the party	4
a_to_b_questions	Detects a person asking questions in a back-and-forth	6
vote_against_party	Metadata: legislator voting against their party	4
pull_quote	Detects a sentence ranked high enough as a pull quote	40

and Appreciation", as they would add the most value to the quote. To label each utterance, we used a neural model that processes both lexical and acoustic features for classification. The model uses a bi-directional LSTM classifier pre-trained on the Switchboard Corpus for lexical clues. The LSTM allows the network to retain the context of the hearing for future tagging. Additionally, the acoustic model uses a CNN to process speech signals which can be equally useful in determining a tag.

The chosen utterances are first ranked based on content using the LexRank algorithm [9], a graph-based centrality scoring of sentences. This is useful as the algorithm can compute the relative importance for each utterance within a document. The algorithm can identify the most central sentences in a cluster that give the greatest amount of information related to the main theme of the document. They are then further ranked by length, sentence root, starting word, bill and geographic mentions, and readability. Sentences are also removed from consideration based on unresolved references and bad structure. After acquiring the highest-ranked pull-quote, the quote is returned along with information about the speaker to be placed in the article.

Back and Forth Argument Detection. The "back and forth" phenom captures news-worthy exchanges and debates that involve two individuals speaking a few lines in rapid interleaving succession. If multiple exchanges appear in a hearing, these exchanges are scored according to average utterance and word length according to the following formula:

$$score = (\alpha * \mu\_all + (1 - \alpha) * \frac{num\_utterances}{hearing\_length}) * \frac{long\_utterances}{num\_utterances} \quad (1)$$

where μ_all is the average utterance length of both speakers and long_utterances is the number of utterances larger than 20 words. The α value (0–1) determines the weights of the average utterance length (for both speakers) and the percentage of the hearing devoted to this exchange. The highest ranking exchange is returned along with some extended personal information about the speakers, which can pulled from the database storage if available.

3.3   Template-Based Text Generation and Planning

bill_mentions	Bill $bill_mentioned1 was also brought up during the bill discussion.|The speakers also brought up $bill_montioned1 throughout the discussion.\|$bill_mentioned1 was referenced in the discussion.
bill_mentions	Bill $bill_mentioned2 was mentioned too.|Another mentioned bill was $bill_mentioned2.|Other bills discussed were $bill_mentioned1 and $billmentioned2.
a_to_b	After the bill was presented a discussion followed, particularly between __$AtoB_person1 and __$AtoB_person2.|During the discussion, a particluar back-and-forth conversation occurred between __$AtoB_person1 and __$AtoB_person2.|The audience was actively debating on the measure afterwards, for example, __$AtoB_person1 and __$AtoB_person2 ended up having a dialogue on this topic.|__$AtoB_person1 and __$AtoB_person2 exchanged a few lines during the discussion.
a_to_b_questions	__$AtoB_questioning_side was mainly asking questions throughout the conversation.\|__$AtoB_questioning_side asked some questions.|__$AtoB_questioning_side wanted some clarifications and information, asking questions.\|__$AtoB_questioning_side asked many good questions.|__$AtoB_questioning_side was clearly engaged in the presentation, asking many questions during the event.\|There were questions from __$AtoB_questioning_side.

Fig. 3. Templates used for text generation of the article.

Phenoms also generate text for the news articles. This text is generated using templates from our library. Each template consists of one or more English sentences with some variable placeholders such as entity names, vote results or bill subjects (see Fig. 3). They are tagged with phenom names, so each phenom can use one or more templates assigned to it. If a template's variables can be entirely resolved using known facts, then the template is used in the article assembly, otherwise, it is discarded. The system tries to find the longest satisfiable template for each phenom. Thus having more information from one phenom can unlock longer templates. All the sentences are then assembled into a final text according to the order produced by the planner.

4   Research Study

Many existing summarization systems are evaluated using metrics [15,17,24, 28] which produce a distance in comparison to a known-good summary. In the case of our study, where there is no gold standard to compare with, the above-mentioned systems are not applicable for evaluation. Therefore, we utilize human evaluations and crowd-source the study.

4.1   Design and Protocol

A questionnaire[1] was designed to determine the effectiveness of the phenom system and assess the quality of the articles produced by such a system.

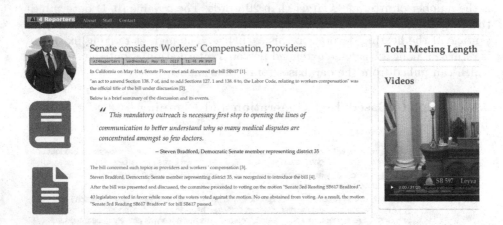

Fig. 4. An example rendered page containing the full article and the video recordings of the hearing that the respondents are presented with during the user study

The study protocol is as follows: the respondent is first presented with a short questionnaire to collect some demographics data, and is requested to watch a video recording of a bill discussion. When they are done watching, another short questionnaire is loaded with questions regarding the newsworthiness of the events happening in the recording. In the next step the respondents are asked to read a summary for the same hearing and answer several questions about the quality of the summary: the article writing style, readability, etc. They are allowed to get back to both the video and the article for more details or comparison.

We use Amazon Mechanical Turk and a questionnaire hosted on Google Forms. The Mechanical Turk service allows researchers to gather many responses quickly, but some will be invalid. For example, the responses that contain incoherent or fake answers. The user study is conducted on two different hearings, chosen as randomly as possible among the ones that in the end provide enough content in the summary to evaluate. The questionnaire given to the Turkers

---

[1] available at https://iatpp.calpoly.edu/slsp2021.

consists of Likert scale and free form text questions, presented to the respondents immediately after they watch the video recording of the hearing and read through the generated article (see Fig. 4).

We collect a data set for each hearing chosen - the first set (Set 1) consists originally of 34 responses, with 22 (64,7%) deemed valid for investigation, while the second test run for another hearing (Set 2) provides 32 answer sets with 20 (62,2%) of them being valid. The invalid answers are selected manually judging mostly by the open text answers - in most of the cases it is incoherent sentences or sentences copied from the article. No specific demographics group was preferred for the user study - only the data concerning the general interest towards legislation news was collected from the respondents. In both answer sets the typical user is a native English speaker consuming state level news data at least weekly.

4.2   Results and Discussion

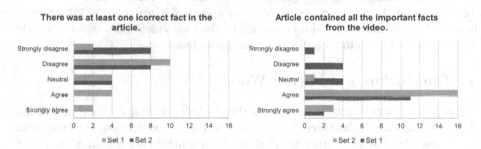

Fig. 5. Statistics collected considering the factual accuracy of the summary produced (N = 22 for Set 1, N = 20 for Set 2)

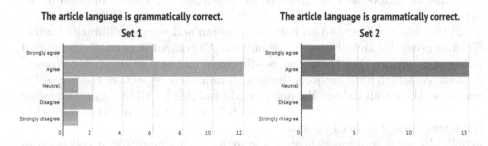

Fig. 6. Statistics collected considering the grammaticality of the summary produced (N = 22 for Set 1, N = 20 for Set 2)

Overall, the articles produced appear to be accurate and correct factually, even though there was a marked difference between the first article and the second. According to the survey, 13 of the 22 respondents in Set 1, and 19 of 20 respondents in Set 2 agree or strongly agree with the statement *"Article contained all the important facts from the video"*. This is contrasted with 6 of 22 respondents in Set 1 and zero of 20 respondents in Set 2, who disagree or strongly disagreeing with the same statement. This is a ratio of 13:6 (Set 1) and 19:0 (Set 2). The same notion was validated by the additional statement *"There was at least one incorrect fact in the article"*, with which 12/22 in Set 1 and 16/20 in Set 2 disagree or strongly disagree. Only 6/22 and 0/22 agree with the same statement for a favorable ratio of 12:6 in Set 1 and 16:0 in Set 2 (see Fig. 5). These findings reveal that the system manages to pick the facts correctly from the transcript, generates articles without omitting expected important events, and does not distort facts.

Positive results are also observed in the grammaticality of the text (see Fig. 6) - an overwhelming majority of the respondents confirm that the summary is a coherent and grammatically correct text. The simplified agree:disagree ratio is 19:3 (Set 1) and 19:0 (Set 2). Only some minor issues were mentioned by some of the respondents in the open text answer fields, such as *"Missing comma before the word "too" at the end of a sentence"*. Judging by the responses to the questions considering the stylistics and readability of the article, most of the Turkers found the article was easy to read, and that it provide the information in an understandable way. More than half of the respondents thought the article was produced by a human and not by an algorithm.

5   Conclusion and Future Work

We propose a solution to help address the declining media coverage in crucial beats such as state legislatures. We design an automatic article generation system using state legislative hearing transcripts. We evaluate the system with a user study. Results indicate our system achieves high performance in accuracy and readability, but doesn't always capture all that is important in the recorded hearings. A sizable minority (9 out of 22) in Set 1 of our study indicated they are either neutral or do not agree that all important facts were captured in the article.

Future work will expand on both our phenom and template libraries leading to more insightful and diverse article output. All content is currently generated from phenoms and thus there's no possibility of any new or surprising types of content. Although most subjects thought all important aspects of the discussion was covered, not all did so. We hope to address this by adding more phenoms to the library and to further design new phenoms based on patterns we find in the transcripts. Another important issue to address is the narrative flow of the article. As some subjects indicated, the article reads more like a collection of sentences and could use internal references and stylistic consistency.

Acknowledgments. The authors thank the John S. and James L. Knight Foundation. The collaboration between Graz University of Technology and Cal Poly was made possible thanks to funding by the Austrian Marshall Plan Foundation. We also thank the Institute for Advanced Technology and Public Policy, and Ms. Christine Robertson for her valuable insights.

References

1. Besley, J.C., Roberts, M.C.: Cuts in newspaper staffs change meeting coverage. Newsp. Res. J. **31**(3), 22–35 (2010)
2. Blakeslee, S., et al.: Digital democracy project: making government more transparent one video at a time. Digit. Hum. **2015** (2015)
3. Budhwar, A., Kuboi, T., Dekhtyar, A., Khosmood, F.: Predicting the vote using legislative speech. In: Proceedings of the 19th Annual International Conference on Digital Government Research: Governance in the Data Age, pp. 1–10 (2018)
4. Bui, T., Frampton, M., Dowding, J., Peters, S.: Extracting decisions from multiparty dialogue using directed graphical models and semantic similarity. In: Proceedings of the SIGDIAL 2009 Conference, pp. 235–243 (2009)
5. Christensen, H., Gotoh, Y., Kolluru, B., Renals, S.: Are extractive text summarisation techniques portable to broadcast news? In: 2003 IEEE Workshop on Automatic Speech Recognition and Understanding (IEEE Cat. No. 03EX721), pp. 489–494. IEEE (2003)
6. Cohen, M.: A new effort boosts a diminished Illinois statehouse press. Columbia Journal. Rev. (2019)
7. Doddington, G.R., Mitchell, A., Przybocki, M.A., Ramshaw, L.A., Strassel, S.M., Weischedel, R.M.: The automatic content extraction (ACE) program-tasks, data, and evaluation. In: Lrec, Lisbon, vol. 2, pp. 837–840 (2004)
8. Enda, J., Matsa, K.E., Boyles, J.L.: America's Shifting Statehouse Press: Can New Players Compensate for Lost Legacy Reporters? Pew Research Center (2014)
9. Erkan, G., Radev, D.R.: LexRank: graph-based lexical centrality as salience in text summarization. J. Artif. Intell. Res. **22**, 457–479 (2004)
10. Feng, X., Feng, X., Qin, B.: A survey on dialogue summarization: Recent advances and new frontiers. arXiv preprint arXiv:2107.03175 (2021)
11. Grieco, E.: Us newsroom employment has dropped by a quarter since 2008, with greatest decline at newspapers. Pew Res. Cent. **9** (2019)
12. Kauffman, D., Khosmood, F., Kuboi, T., Dekhtyar, A.: Learning alignments from legislative discourse. In: Proceedings of the 19th Annual International Conference on Digital Government Research: Governance in the Data Age, pp. 1–2 (2018)
13. Leppänen, L., Munezero, M., Granroth-Wilding, M., Toivonen, H.: Data-driven news generation for automated journalism. In: Proceedings of the 10th International Conference on Natural Language Generation, pp. 188–197 (2017)
14. Li, Q., et al.: Reinforcement learning-based dialogue guided event extraction to exploit argument relations. arXiv preprint arXiv:2106.12384 (2021)
15. Lin, C.Y.: ROUGE: a package for automatic evaluation of summaries. In: Text Summarization Branches Out, pp. 74–81 (2004)
16. Liu, F., Liu, Y.: From extractive to abstractive meeting summaries: can it be done by sentence compression? In: Proceedings of the ACL-IJCNLP 2009 Conference Short Papers, pp. 261–264 (2009)
17. Mani, I., Maybury, M.T.: Automatic Summarization. J. Benjamins Publishing Company, New York (2001)
18. Marum, A.: Oregon's dwindling statehouse reporters are 'treading water'. Columbia Journal. Rev. (2018)
19. Murray, G., Carenini, G., Ng, R.: Generating and validating abstracts of meeting conversations: a user study. In: Proceedings of the 6th International Natural Language Generation Conference (2010)

20. Nesterenko, L.: Building a system for stock news generation in Russian. In: Proceedings of the 2nd International Workshop on Natural Language Generation and the Semantic Web (WebNLG 2016), pp. 37–40 (2016)

21. Nguyen, T.M., Nguyen, T.H.: One for all: neural joint modeling of entities and events. In: Proceedings of the AAAI Conference on Artificial Intelligence, vol. 33, pp. 6851–6858 (2019)

22. Ortega, D., Vu, N.T.: Lexico-acoustic neural-based models for dialog act classification. In: 2018 IEEE International Conference on Acoustics, Speech and Signal Processing (ICASSP), pp. 6194–6198. IEEE (2018)

23. Oya, T., Mehdad, Y., Carenini, G., Ng, R.: A template-based abstractive meeting summarization: leveraging summary and source text relationships. In: Proceedings of the 8th International Natural Language Generation Conference (INLG), pp. 45–53 (2014)

24. Papineni, K., Roukos, S., Ward, T., Zhu, W.J.: BLEU: a method for automatic evaluation of machine translation. In: Proceedings of the 40th Annual Meeting of the Association for Computational Linguistics, pp. 311–318 (2002)

25. Riedhammer, K., Favre, B., Hakkani-Tür, D.: Long story short-global unsupervised models for keyphrase based meeting summarization. Speech Commun. **52**(10), 801–815 (2010)

26. Ruprechter, T., Khosmood, F., Gütl, C.: Deconstructing human-assisted video transcription and annotation for legislative proceedings. Digit. Govern.: Res. Pract. **1**(3), 1–24 (2020)

27. Ruprechter, T., Khosmood, F., Kuboi, T., Dekhtyar, A., Gütl, C.: Gaining efficiency in human assisted transcription and speech annotation in legislative proceedings. In: Proceedings of the 19th Annual International Conference on Digital Government Research: Governance in the Data Age, pp. 1–2 (2018)

28. Saggion, H., Radev, D., Teufel, S., Lam, W.: Meta-evaluation of summaries in a cross-lingual environment using content-based metrics. In: COLING 2002: The 19th International Conference on Computational Linguistics (2002)

29. Schonfeld, E.: Automated news comes to sports coverage via statsheet. TechCrunch, 12 November 2010 (2010)

30. Shaw, A.: As statehouse press corps dwindles, other reliable news sources needed. Better Government Association (2017)

31. Stolcke, A., et al.: Dialogue act modeling for automatic tagging and recognition of conversational speech. Comput. Linguist. **26**(3), 339–373 (2000)

32. Wang, L., Cardie, C.: Domain-independent abstract generation for focused meeting summarization. In: Proceedings of the 51st Annual Meeting of the Association for Computational Linguistics (Volume 1: Long Papers), pp. 1395–1405 (2013)

33. Weiss, S.: 140 characters of news. State Legislatures **41** (2015)

34. Williams, A.T.: Measuring the journalism crisis: Developing new approaches that help the public connect to the issue. Int. J. Commun. (19328036) **11** (2017)

35. Xie, S., Liu, Y., Lin, H.: Evaluating the effectiveness of features and sampling in extractive meeting summarization. In: 2008 IEEE Spoken Language Technology Workshop, pp. 157–160. IEEE (2008)

Comparison of Czech Transformers on Text Classification Tasks

Jan Lehečka(✉) and Jan Švec

Department of Cybernetics, University of West Bohemia in Pilsen,
Univerzitní 2732/8, 301 00 Pilsen, Czech Republic
{jlehecka,honzas}@kky.zcu.cz

Abstract. In this paper, we present our progress in pre-training monolingual Transformers for Czech and contribute to the research community
by releasing our models for public. The need for such models emerged
from our effort to employ Transformers in our language-specific tasks,
but we found the performance of the published multilingual models to be
very limited. Since the multilingual models are usually pre-trained from
100+ languages, most of low-resourced languages (including Czech) are
under-represented in these models. At the same time, there is a huge
amount of monolingual training data available in web archives like Common Crawl. We have pre-trained and publicly released two monolingual
Czech Transformers and compared them with relevant public models,
trained (at least partially) for Czech. The paper presents the Transformers pre-training procedure as well as a comparison of pre-trained models
on text classification task from various domains.

Keywords: Text categorization and summarization · Monolingual
transformers · Sentiment analysis · Multi-label topic identification

1 Introduction

In the last few years, deep neural networks based on Transformers [4] have dom-
inated the research field of Natural Language Processing (NLP) and Natural
Language Understanding (NLU). Self-attention [19] Transformers and especially
the self-supervised-trained variants known as BERT (Bidirectional Encoder Representations from Transformers) models, have achieved amazing results in many
tasks, including text classification [1,9,17].

Because the self-supervised pre-training of Transformers is computationally
very costly, researchers around the world publish their pre-trained models in
order to compete on a large variety of NLP and NLU tasks, however the majority
of these models are monolingual English models or multilingual models including
many languages at once. Since the multilingual models are usually pre-trained
from Wikipedia dumps or large web archives, most of low-resourced languages
are under-represented in these models. This makes practical use of such models
in non-English languages very limited. Therefore in present days, we observe

© Springer Nature Switzerland AG 2021
L. Espinosa-Anke et al. (Eds.): SLSP 2021, LNAI 13062, pp. 27–37, 2021.
https://doi.org/10.1007/978-3-030-89579-2_3

a significant trend in pre-training also monolingual transformers for other languages, e.g. CamemBERT for French [13], Finnish BERT [20] and very recently also Czech models Czert [15] and RobeCzech [16].

The trend in pre-training monolingual Transformers is boosted by the accessibility of powerful hardware devices (especially graphical and text processing units, GPUs and TPUs) and by the availability of huge text corpora even for languages with relatively small number of native speakers. For example, many researchers use data from the Common Crawl project[1] which is a huge public web archive consisting of petabytes of crawled web pages.

In this paper, we present our effort to pre-train two monolingual Czech Transformers (BERT and RoBERTa) from different datasets and we compare the performance with published multilingual and monolingual Czech models on downstream text classification tasks. We release both our pre-trained models publicly on HuggingFace hub[2].

2   Related Work

Aside from the famous English BERT-base model [4], also its multilingual variant has been published by Google researchers. It has been trained from Wikipedia dumps of 104 languages, including Czech. We denote this model *MultiBERT* in this paper.

BERT model trained from Wikipedia dumps of only four Slavic languages (Russian, Bulgarian, Czech and Polish) has been presented in [2]. The model was not pre-trained from the scratch, but MultiBERT weights were used to initialize the model. We denote this model *SlavicBERT* in this paper.

Researchers from Facebook have published multilingual *XLM-RoBERTa* model [3] pre-trained on one hundred languages (including Czech), using more than two terabytes of filtered Common Crawl data. Two sizes of the model have been published: base model with 270 million parameters and a large one with 550 million parameters.

There are also two recent papers presenting monolingual Czech models. *Czert* [15] is a BERT model pre-trained from a mixture of Czech national corpus [7], Wikipedia pages and self-crawled news dataset. The second one, *RobeCzech* [16], is a RoBERTa [11] model pre-trained from Czech national corpus [7], Wikipedia pages, a collection of Czech newspaper and magazine articles and a Czech part of the web corpus W2C [12]. However, these models were pre-trained from a standard (literary) Czech, which makes them unsuitable for tasks where texts are often written in a common (spoken) Czech, such as social media posts or product reviews. We provide the basic information about described Transformers in the Table 2.

---

[1] https://commoncrawl.org.
[2] https://huggingface.co/fav-kky.

3 Datasets

3.1 Pre-training Datasets

Since self-supervised pre-training of Transformers requires a huge amount of unlabeled text, our aim was to get as much cleaned Czech text as possible. In the following paragraphs, we provide a description of datasets we were experimenting with.

News Corpus. We used Czech news dataset from our web-mining framework [18]. The dataset consists mainly of web-crawled news articles and transcripts of selected TV and radio shows. In this dataset, we cleaned boilerplate and HTML tags from original documents and thoroughly processed each document with normalization, true-casing and rule-based word substitution dealing with typos, multiwords and word-form unification. See [18] for more details. The corpus contains 3.3 billion words (20.5 GB of cleaned text).

C5. Czech Colossal Clean Crawled Corpus (C5) generated from the Common Crawl project is a Czech mutation of the English C4 dataset [14]. We used the language information provided in the index files to select Czech records only. The corresponding plain texts (stored in WET archives) were downloaded. To clean the data, we followed almost the same rules which were used to pre-process the C4 dataset, i.e.:

- We only retained lines that ended with a terminal punctuation mark ("."," "?" or "!").
- We removed lines containing "javascript" or "cookies".
- We removed web pages containing offensive words or strings "lorem ipsum" or "{".
- We retained only pages classified as Czech with probability of at least 0.99 according to `langdetect`[3] tool.
- To deduplicate the dataset, we discarded all but one line occurring more than once in the data set.
- We only retained lines with at least 3 words and pages with at least 5 sentences.

This simple yet rigorous cleaning process removed about 98% of downloaded plain texts from the dataset, mainly due to the deduplication (the more data we downloaded, the harder it was to find a new unobserved line).

We downloaded and cleaned recent crawls up to the crawl from August 2018, which is the first one providing the language column in its index. Together, we processed 25 crawls (from August 2018 to October 2020) and the resulted dataset contains almost 13 billion words (93 GB of cleaned text[4]).

---

[3] https://pypi.org/project/langdetect/.

[4] To get a picture how huge the Common Crawl archive is, consider that this 93 GB of text is (due to rigorous cleaning rules) just about 2% of all Czech texts we downloaded and Czech records occupy just about 1% of the full archive.

3.2   Text Classification Datasets

We decided to fine-tune and evaluate Transformers on a text classification task. Specifically, we fine-tuned models for sentiment analysis and multi-label topic identification tasks as these tasks are in our main focus. To keep our experiments reproducible, we experimented mainly with publicly available datasets. In the following paragraphs, we describe five Czech datasets used in the experiments together with the basic statistics of the datasets in Table 1. The first three datasets are for sentiment analysis task. All of them were created in [5] and contain text samples with positive, neutral or negative class labels. The last two datasets are for multi-label topic identification task. Both of them are from news domain and contain crawled news articles, each with at least one topic label.

Table 1. Czech text classification datasets statistics. The table shows numbers of documents in train, development and test datasets, number of classes and the label cardinality (i.e. the average number of labels per document) of the datasets.

Dataset	Train	Devel	Test	Classes	LabCard
CSFD	91 381	–	–	3	1
MALL	145 307	–	–	3	1
FCB	9 752	–	–	3	1
CN	184 313	20 479	43 762	577	3.06
CTDC	11 955	2 538	–	37	2.55

CSFD. Movie review dataset consists of 91 thousand movie reviews from the Czech-Slovak Movie Database (ČSFD)[5]. Based on movie ratings, each review was classified into positive, neutral or negative sentiment class.

MALL. Product review dataset consists of 145 thousand posts from a large Czech e-shop Mall.cz[6] which offers a wide range of products. Similar to CSFD dataset, each post was classified into positive, neutral or negative sentiment class based on the user's ratings.

FCB. The Facebook dataset contains 10 thousand posts selected from several Facebook pages with a large Czech fan base. The sentiment of individual posts was assigned manually by multiple annotators. We ignored the 248 bipolar posts and trained models only with three sentiment classes as suggested by authors of the dataset [5].

---

[5] https://www.csfd.cz.
[6] https://www.mall.cz.

CTDC. Czech Text Document Corpus (CTDC)[7] is a news dataset described in [6]. The dataset consists of 12 thousand news articles annotated with category labels. In order to keep our results comparable with other papers (such as [10, 15]), we used the same setup and selected only 37 (out of 60) most frequent labels.

CN. Many news servers publish articles along with topic labels assigned by journalists. We keep this information in our framework [18] and map the most frequent labels into a standardized IPTC news topic tree[8]. We created a CN dataset by selecting articles with assigned IPTC topic labels from one news server (`ceskenoviny.cz`, hence the name CN) which we believe to be labeled thoroughly and consistently. The dataset contains 250 thousand articles from 10-years epoch. Each article has assigned one or more topic labels. The total number of labels in the dataset is 577. Unfortunately, we do not have licence to publish this dataset. It is, however, the only private dataset appearing in this paper.

4   Models

In this paper, we present two new Transformers. We named our models Flexible Embedding Representation NETwork (FERNET). Our aim was to pre-train one general model suitable for a wide range of domains and one single-domain model specifically for the news domain, which is in our main focus. After we had pre-trained the first BERT model, we reflected the change of paradigm and switched to the RoBERTa architecture when pre-training the second one. We tabulate basic summary of our models along with all other models we have experimented with in Table 2, and describe them in the following paragraphs.

FERNET-C5. This is a BERT model trained from scratch using the C5 dataset. The model has the same architecture as BERT-base model [4], i.e. 12 transformation blocks, 12 attention heads and the hidden size of 768 neurons.

In contrast to Google's BERT models, we used SentencePiece tokenization [8] instead of the Google's internal WordPiece tokenization. We trained a BPE SentencePiece model from the underlying dataset with vocabulary size set to 100 thousands. Since the datasets contains a small portion of non-Czech text fragments, we carefully tuned the character coverage parameter to fully cover all Czech characters and only reasonable number of graphemes from other languages. We kept original casing of the dataset.

FERNET-News. This is a RoBERTa model trained from scratch using our thoroughly pre-processed News dataset. The model has the same architecture as

---

[7] http://ctdc.kiv.zcu.cz.
[8] http://cv.iptc.org/newscodes/mediatopic.

Table 2. Summary of pre-trained Transformers trained at least partially on Czech. The first four models are multilingual, the next two models are already published monolingual Czech Transformers. The two models at the bottom of the table are our new models presented in this paper. The sizes of vocabularies are in thousands (k) and number of trainable parameters are in millions (M). Reported numbers of parameters include language model head on top of the model. SPM stands for SentencePiece tokenization [8], BBPE for Byte-level Byte-Pair Encoding and CC for Common Crawl.

Model	Architecture	Tokenization	Vocab	Pre-train data	Params
MultiBERT [4]	BERT-base	WordPiece	120 k	Wiki, 104 langs	179 M
SlavicBERT [2]	BERT-base	subword-nmt	120 k	Wiki, 4 langs	179 M
XLM-RoBERTa-base [3]	RoBERTa-base	SPM	250 k	CC, 100 langs (2 TB)	278 M
XLM-RoBERTa-large [3]	RoBERTa-large	SPM	250 k	CC, 100 langs (2 TB)	560 M
Czert [15]	BERT-base	WordPiece	40 k	Nat+Wiki+News (37 GB)	110 M
RobeCzech [16]	RoBERTa-base	BBPE	52 k	Nat+Wiki+Czes+W2C	126 M
FERNET-C5	BERT-base	SPM	100 k	C5 (93 GB)	164 M
FERNET-News	RoBERTa-base	BBPE	50 k	News Corpus (21 GB)	124 M

RoBERTa-base model [11], i.e. 12 transformation blocks, 12 attention heads and the hidden size of 768 neurons. To tokenize input texts, we trained byte-level BPE tokenizer with the vocabulary size 50 000 tokens, as suggested in the paper.

4.1 Pre-training

Our BERT model was pre-trained using two standard tasks: masked language modeling (MLM), which is a fill-in-the-blank task, where the model is taught to predict the masked words, and next sentence prediction (NSP) task, which is the classification task of distinguishing between inputs with two consequent sentences and inputs with two randomly chosen sentences from a corpus. We pre-trained the model for 6.5 million gradient steps in total (3.5 M steps with 128-long inputs and batch size 256, and 3 M steps with 512-long inputs and batch size 128). For both input variants, we used whole word masking, duplication factor of 5 and the learning rate warmed up over the first 100 000 steps to a peak value at 1×10^{-4}. To pre-train the model, we used the software provided by Google researchers[9] and to speed up the training, we used one 8-core TPU with 128 GB of memory. The pre-training took about two months.

Our RoBERTa model was pre-trained using MLM task only. We trained the model for 600 thousand gradient steps with batch size 2048: first 500 thousand steps with 128-long inputs, 24 000 warmup steps and learning rate 4×10^{-4}, and the following 100 thousand steps with 512-long inputs, 2 000 warmup steps and learning rate 1×10^{-4}. Since the news dataset is rather small (but very clean), the model iterated about 40-times over the whole training corpus during the pre-training. However, due to dynamic language model masking, the model saw differently masked input samples in each epoch. To pre-train the model, we

---

[9] https://github.com/google-research/bert.

used the HuggingFace's Transformers software[10] and to speed up the training, we used two Nvidia A100 GPUs. The pre-training took about one month.

4.2 Fine-Tuning

We run fine-tuning experiments on single GPUs using HuggingFace's Transformers software.

To fine-tune models for sentiment analysis, we fixed the peak learning rate to 1×10^{-5}, maximum sequence length 128 tokens, batch size 64, 10 training epochs and 100 warmup steps. We were also experimenting with other hyperparameters setting, but this one gave consistently the best results across all models and datasets. After each training epoch, we evaluated the model on the development dataset (if not present in the dataset, we randomly held out 10% of the training samples as a development data) and at the end, we used the best model for evaluation on the test dataset.

To fine-tune models for multi-label topic identification task, we fixed the peak learning rate to 5×10^{-5}, maximum sequence length 512 tokens and batch size 64. We trained on the CTDC dataset for 20 epochs whereas to train on CN dataset, 10 epochs was enough, because the dataset is much larger.

4.3 Evaluation

For both BERT and RoBERTa model architectures, we predicted classes using standard pooling layers for sequence classification implemented in the Transformers library along with the corresponding model types. This pooling layer uses the last hidden state of the first input token ([CLS] or <s>) to make predictions.

In the case of sentiment analysis task, we assigned sentiment class with the highest predicted value for each document and evaluated F1 score.

In the case of multi-label text classification tasks, we added a sigmoid activation function to the final output layer of each model, thus the output for each document was a vector of per-label scores $s = (s_1, s_2, ..., s_K)$, $s_i \in (0, 1)$, where K is total number of labels. We converted this "soft" predictions into binary "hard" predictions by following thresholding strategy: assign i-th label, if

$$\frac{s_i}{\max(s_1, s_2, ..., s_K)} \geq 0.5. \tag{1}$$

This strategy gave consistently better results than simple thresholding sigmoid outputs on 0.5. After thresholding, we evaluated sample-averaged F1 score.

For datasets which do not contain separated test dataset for evaluation (all but CN), we performed 10-fold cross-validation. For sentiment analysis datasets, we split the data while preserving the percentage of samples for each class (i.e. stratified cross-validation), for topic identification datasets, we split the data

---

[10] https://huggingface.co/transformers.

randomly. For each pre-trained model and each dataset, we fine-tuned 10 different models (one for each fold), averaged the results and reported the mean and standard deviation.

5   Results

We summarize the experimental results in Table 3 (sentiment analysis task) and Table 4 (multi-label topic identification task). In both tables, we first report "base-model-size" category results as these models have more or less comparable number of training parameters (see Table 2 for details). In the last row, we report also results with XLM-RoBERTa-large model, which has, however, several-times more parameters.

Table 3. Table of results for sentiment analysis task. We report comparison of Transformers in the mean of F1 score on three Czech datasets. Aside from base-sized models, we also report comparison with XLM-RoBERTa-large, which is a much larger model.

Model	CSFD	MALL	FCB
MultiBERT [4]	81.04 (±0.31)	76.91 (±0.47)	73.24 (±2.37)
SlavicBERT [2]	82.14 (±0.33)	78.17 (±0.39)	74.19 (±1.85)
XLM-RoBERTa-base [3]	83.74 (±0.39)	77.03 (±0.61)	78.92 (±1.71)
Czert [15]	83.76 (±0.32)	79.35 (±0.52)	78.34 (±1.59)
RobeCzech [16]	85.01 (±0.43)	78.18 (±0.36)	79.12 (±1.39)
FERNET-C5	**85.36** (±0.30)	**79.75** (±0.37)	**81.07** (±1.63)
FERNET-News	84.21 (±0.41)	77.43 (±0.49)	75.94 (±1.85)
XLM-RoBERTa-large [3]	86.03 (±0.32)	79.94 (±0.48)	82.23 (±1.29)

From the sentiment analysis results (Table 3), we can see that among base-sized models, our FERNET-C5 model scored best across all datasets. We attribute this superiority to the underlying pre-training data of the model. In all three fine-tuning datasets, we are dealing with text samples written by common people, often under a strong emotions (e.g. angry about a defective product they just bought, delighted about a movie they just saw etc.). It is also a frequent phenomenon (especially in social media domain), that people are writing posts in a common (spoken) Czech, which is extraordinarily different from literary (standard) Czech, used e.g. by journalists. Moreover, diacritical and punctuation marks are often missing in user's posts. Since the C5 dataset was crawled from miscellaneous Czech web pages, all mentioned phenomenons are covered richly in this dataset. Other popular pre-training data sources (Wikipedia, news articles, books etc.) contain mainly standard Czech, which makes models pre-trained from such sources less suitable for this task. Good results scored by multilingual XLM-RoBERTa models, which were trained from Common Crawl data as well, support this hypothesis.

Table 4. Table of results for multi-label topic identification task. We report comparison of Transformers in the mean of sample-averaged F1 score on two Czech datasets. Aside from base-sized models, we also report comparison with XLM-RoBERTa-large, which is a much larger model.

Model	CTDC	CN
MultiBERT [4]	89.07 ($\pm$ 0.63)	79.83
SlavicBERT [2]	89.59 ($\pm$ 0.48)	80.46
XLM-RoBERTa-base [3]	88.76 ($\pm$ 0.63)	80.35
Czert [15]	90.23 ($\pm$ 0.49)	81.50
RobeCzech [16]	90.47 ($\pm$ 0.53)	81.20
FERNET-C5	**91.25** ($\pm$ 0.38)	82.13
FERNET-News	90.85 ($\pm$ 0.47)	**82.29**
XLM-RoBERTa-large [3]	91.18 ($\pm$ 0.53)	82.78

From the multi-label topic identification results (Table 4), we can see that both our models scored better than other base-sized models. Since our RoBERTa model was pre-trained from in-domain data for this task (news) and RoBERTa is known to perform better than BERT models, we expected the FERNET-News model to score significantly better than other models. However, we were surprised by the universality of the FERNET-C5 model, which performed comparably well, and on the CTDC dataset, the FERNET-C5 model even outperformed in-domain FERNET-News model. Moreover, FERNET-C5 outperformed even the 5-times larger XML-RoBERTa-large model, however the difference is not statistically significant. We attribute such good results of the FERNET-C5 model to the significantly larger pre-training dataset.

To summarize our results, our models outperformed all public multilingual and monolingual Transformers in the base-model-size category on the text classification tasks. The model pre-trained from 93 GB of filtered Czech Common Crawl dataset (FERNET-C5) demonstrated surprising universality across all domains we were experimenting with. Results achieved by our base-sized models are also comparable (although slightly worse in most cases) with results scored by 5-times larger model XLM-RoBERTa-large.

6   Conclusions

In this paper, we used large Czech text datasets to pre-train two monolingual Transformers, one from 93 GB of filtered Czech Common Crawl dataset (FERNET-C5) and one from 21 GB of thoroughly cleaned self-crawled news corpus (FERNET-News). The paper describes the dataset preprocessing as well as the models pre-training procedure and evaluates the models on five different Czech text-classification tasks. Our models outperformed all published multilingual and monolingual Transformers in the base-model-size category. The model pre-trained from Common Crawl (FERNET-C5) demonstrated surprising universality across all domains we were experimenting with.

Since Czech is a special language due to its significant differences between spoken and written form, and since internet users often tend to use spoken form with many grammatical errors to write down a content to be classified (social media posts, reviews etc.), it could be very beneficial to cover these phenomena also in the pre-training corpus. Our results showed that model trained from Common Crawl has universal usage across all domains, whereas models pre-trained from popular literary Czech datasets like Wikipedia or news articles, has limited usage for literary datasets only. For example, our FERNET-News pre-trained from thoroughly cleaned news corpus failed to classify correct sentiment of Facebook posts (F1 75.94 vs. 81.07).

Our models are publicly available for research purposes on HuggingFace hub (See footnote 2).

Acknowledgments. This research was supported by the Ministry of Culture of the Czech Republic, project No. DG18P02OVV016.

References

1. Adhikari, A., Ram, A., Tang, R., Lin, J.: DocBERT: BERT for document classification. arXiv preprint arXiv:1904.08398 (2019)
2. Arkhipov, M., Trofimova, M., Kuratov, Y., Sorokin, A.: Tuning multilingual transformers for language-specific named entity recognition. In: Proceedings of the 7th Workshop on Balto-Slavic Natural Language Processing, pp. 89–93 (2019)
3. Conneau, A., et al.: Unsupervised cross-lingual representation learning at scale. In: Proceedings of the 58th Annual Meeting of the Association for Computational Linguistics, pp. 8440–8451 (2020)
4. Devlin, J., Chang, M.W., Lee, K., Toutanova, K.: BERT: pre-training of deep bidirectional transformers for language understanding. In: Proceedings of the 2019 Conference of the North American Chapter of the Association for Computational Linguistics: Human Language Technologies, Volume 1 (Long and Short Papers), pp. 4171–4186. Association for Computational Linguistics, Minneapolis (2019)
5. Habernal, I., Ptáček, T., Steinberger, J.: Sentiment analysis in Czech social media using supervised machine learning. In: Proceedings of the 4th Workshop on Computational Approaches to Subjectivity, Sentiment and Social Media Analysis, pp. 65–74 (2013)
6. Kral, P., Lenc, L.: Czech text document corpus v 2.0. In: Calzolari, N., et al. (eds.) Proceedings of the Eleventh International Conference on Language Resources and Evaluation (LREC 2018). European Language Resources Association (ELRA), Paris, France, May 2018
7. Křen, M., et al.: SYN v4: large corpus of written Czech (2016). http://hdl.handle.net/11234/1-1846. LINDAT/CLARIAH-CZ digital library at the Institute of Formal and Applied Linguistics (ÚFAL), Faculty of Mathematics and Physics, Charles University
8. Kudo, T., Richardson, J.: SentencePiece: a simple and language independent subword tokenizer and detokenizer for neural text processing. In: Proceedings of the 2018 Conference on Empirical Methods in Natural Language Processing: System Demonstrations, pp. 66–71. Association for Computational Linguistics, Brussels, November 2018. https://doi.org/10.18653/v1/D18-2012. https://www.aclweb.org/anthology/D18-2012

9. Lehečka, J., Švec, J., Ircing, P., Šmídl, L.: Adjusting BERT's pooling layer for large-scale multi-label text classification. In: Sojka, P., Kopeček, I., Pala, K., Horák, A. (eds.) TSD 2020. LNCS (LNAI), vol. 12284, pp. 214–221. Springer, Cham (2020). https://doi.org/10.1007/978-3-030-58323-1_23

10. Lenc, L., Král, P.: Deep neural networks for Czech multi-label document classification. In: Gelbukh, A. (ed.) CICLing 2016. LNCS, vol. 9624, pp. 460–471. Springer, Cham (2018). https://doi.org/10.1007/978-3-319-75487-1_36

11. Liu, Y., et al.: RoBERTa: a robustly optimized BERT pretraining approach. arXiv preprint arXiv:1907.11692 (2019)

12. Majliš, M.: W2C - web to corpus - corpora (2011). http://hdl.handle.net/11858/00-097C-0000-0022-6133-9. LINDAT/CLARIAH-CZ digital library at the Institute of Formal and Applied Linguistics (ÚFAL), Faculty of Mathematics and Physics, Charles University

13. Martin, L., et al.: CamemBERT: a tasty French language model. In: Proceedings of the 58th Annual Meeting of the Association for Computational Linguistics, pp. 7203–7219. Association for Computational Linguistics, July 2020. https://doi.org/10.18653/v1/2020.acl-main.645. https://www.aclweb.org/anthology/2020.acl-main.645

14. Raffel, C., et al.: Exploring the limits of transfer learning with a unified text-to-text transformer. J. Mach. Learn. Res. **21**(140), 1–67 (2020). http://jmlr.org/papers/v21/20-074.html

15. Sido, J., Pražák, O., Přibáň, P., Pašek, J., Seják, M., Konopík, M.: Czert-Czech BERT-like model for language representation. arXiv preprint arXiv:2103.13031 (2021)

16. Straka, M., Náplava, J., Straková, J., Samuel, D.: RobeCzech: Czech RoBERTa, a monolingual contextualized language representation model. arXiv preprint arXiv:2105.11314 (2021)

17. Sun, C., Qiu, X., Xu, Y., Huang, X.: How to fine-tune BERT for text classification? In: Sun, M., Huang, X., Ji, H., Liu, Z., Liu, Y. (eds.) CCL 2019. LNCS (LNAI), vol. 11856, pp. 194–206. Springer, Cham (2019). https://doi.org/10.1007/978-3-030-32381-3_16

18. Švec, J., et al.: General framework for mining, processing and storing large amounts of electronic texts for language modeling purposes. Lang. Resour. Eval. **48**(2), 227–248 (2014)

19. Vaswani, A., et al.: Attention is all you need. In: Advances in Neural Information Processing Systems, pp. 5998–6008 (2017)

20. Virtanen, A., et al.: Multilingual is not enough: BERT for finnish. arXiv preprint arXiv:1912.07076 (2019)

Constructing Sentiment Lexicon
with Game for Annotation Collection

Lukáš Radoský[(✉)] and Miroslav Blšták

Kempelen Institute of Intelligent Technologies, Mlynské nivy 5,
811 09 Ružinov, Slovakia
{lukas.radosky,miroslav.blstak}@kinit.sk
https://kinit.sk/

Abstract. While research of sentiment analysis became very popular
on the global scope, in Slovak language as an under-resourced language
there are still many issues to be tackled, especially the lack of resources.
In this paper, we introduce a sentiment analysis game designed to collect sentiment annotations. The game is intended for a single player
who, motivated by game score, chooses the sentiment category of each
word of a sentence. We describe the annotation collection process during
which over 12 500 annotations of individual words and over 1 000 annotations of entire sentences were obtained within a week. The collected
annotations were used to construct a sentiment lexicon. Using artificial bee colony algorithm, optimal lexicon construction parameters were
discovered. To evaluate the final lexicon's usefulness, we applied simple sentence-level lexicon-based sentiment analysis methods on a manually annotated dataset of mobile phone reviews. The same was done
with other existing lexicons for comparison. The results of our experiments show that collecting annotations using our game can be useful as
a method for constructing a sentiment lexicon.

Keywords: Lexicons and dictionaries · Sentiment analysis ·
Crowdsourcing · Games with a purpose · Natural language processing

1 Introduction

Sentiment analysis is a field of study analyzing emotional attitudes (sentiments)
that people express towards concepts or their specific aspects [11]. The simplest
sentiment consists of polarity and intensity. Polarity defines whether the attitude
is positive or negative [15]. Neutral polarity can also be introduced to represent
absence of sentiment [11]. Intensity defines strength of the attitude on a defined
scale. Sentiment can also be nominal measure with values as *anger*, *sadness*,
happiness etc. [4]. Sentiment can be expressed textually, visually or auditorially,
but sentiment analysis is focused mostly on textually captured sentiment [21].

Automated methods of sentiment analysis allow utilization of opinions
expressed by people online. Companies and governments use this approach to
analyse how the public perceives their products or activities. For example, if

© Springer Nature Switzerland AG 2021
L. Espinosa-Anke et al. (Eds.): SLSP 2021, LNAI 13062, pp. 38–49, 2021.
https://doi.org/10.1007/978-3-030-89579-2_4

most reviews of a product carry negative sentiment, it indicates that the company should make adjustments to the product.

Sentiment analysis inherits problems of related research fields. Many words have ambiguous sentiment unless their context is considered [11], e.g. 'long waiting' bears negative while 'long battery charge duration' bears positive sentiment. Many machine-learning-based methods suffer from lack of interpretability - it is not easy to determine why the model predicted the given value [17]. Training dataset limitations cause bias in model predictions, e.g. models used by American justice system are proven to be more likely to falsely label a black person as high-risk and a white person as low-risk [3]. In sentiment analysis, Mexican restaurants may receive lower ratings due to negative associations with Mexican refugees [19]. Lexicon-based methods are quite transparent, but they only work with words contained by the used lexicon.

Resources are numerous for major languages such as English, but scarce for minor languages such as Slovak. We utilized crowdsourcing approach to collect sentiment annotations. The annotation process was concealed as a game with a purpose (GWAP), even if in a simple manner. Using collected annotations, we have built a Slovak sentiment lexicon, and evaluated its usefulness by applying simple lexicon-based methods on a manually annotated dataset.

We overview some existing GWAPs and previously published work related to sentiment analysis in Slovak in Sect. 2. Our GWAP is introduced in Sect. 3. In Sect. 4, we look in detail at the collected data, lexicon construction, its evaluation, and comparison towards other lexicons. Section 5 briefly concludes this paper.

2   Related Work

2.1   Games with a Purpose

Effectiveness of both machine learning approach and lexicon-based approach [11] strongly depends on quality of available annotations. High-quality annotations are created by domain experts [18]. Collecting such annotations is not scalable, as there are usually not many domain experts available, making the process slow, as admitted by Alqaryouti et al. [2]. Experts usually need to be paid, making it also expensive [18].

Bučar et al. [5] collected sentiment annotations for over 10 000 news documents in Slovene from domain experts. In their effort, all pitfalls of expert annotations manifested. Six annotators had to be trained and paid, and annotation collection took a year. The benefit was guaranteed annotation quality.

Another approach is to collect data from common internet users on a large scale. This provides scalability at the cost of annotation quality, which needs to be tackled very carefully. Crowdsourcing is particularly useful when working with minor languages or specific domains. It is important to consider that collected annotations do not need to be an attempt to estimate a grand-truth value objectively linked to the subject - annotations may represent the person's own perception.

Some GWAPs conceal their true purpose, providing entertaining or useful functionality on top of it. In *The ESP game* [1] one player describes an image with a single word, the other player is supposed to guess that word. The activity itself is entertaining and provides useful output - image labels. In 4 months, they collected almost 1.3 million labels. *PexAce* [23] hides its purpose particularly well. A single player is supposed to find pairs of matching cards by flipping them. The game allows to make notes about pictures on the cards, making guessing easier. The notes are used as image labels. Over 22 000 labels were collected.

Sometimes it is difficult to conceal the purpose of a GWAP. In such cases, GWAPs employ gamification elements. In *Guesstiment* [13], one player determines sentiment of sentence, then another player guesses which word contributed the most to the selected sentiment. While the game made the activity entertaining, it did not conceal its purpose. 697 annotations was collected. In *Cipher* [20], a single player is supposed to detect spelling errors in text. The game's strong point is its medieval-like visual design. In 2 weeks, they collected over 4 500 pieces of annotation data.

2.2   Sentiment Analysis in Slovak

Okruhlica [14] created Slovak sentiment lexicon using Hatzivassiloglou-McKeown induction algorithm [8], starting with a small sample of manually annotated data. The idea of the algorithm is that if words are grouped in a certain syntactical construction, their sentiment is similar, e.g. if sentiment of 'nice' is known to be positive, sentence 'He was nice and charming' indicates that word 'charming' also has positive sentiment. Resulting sentiment lexicon[1] contains 3018 adjectives and 3018 adverbs with 7 sentiment classes.

Mikula and Machová [12] created a Slovak sentiment lexicon by optimizing the sentiment of words using particle swarm optimization (PSO). Fitness function of lexicon was chosen metric of 2-class classification on translated English dataset. Their results outperformed human annotators, with F1-score crossing 85% in some cases.

Pecár et al. [16] researched efficiency of LSTM neural networks combined with word embeddings for sentiment analysis in Slovak. On two datasets - customer reviews and tweets - they outperformed previous approaches to sentiment analysis in Slovak, in some cases reaching F1-score over 80% with 3-class classification.

Krchňavý and Šimko [10] applied machine learning sentiment analysis methods on document-level in Slovak language. They were rather successful in binary classification, achieving accuracy over 82% with Maximum Entropy Classifier, which however drops to cca 41% with 5-class classification.

To sum up, sentiment analysis of Slovak has already received some attention. All previously used approaches, i.e. lexicon-based [12], machine-learning-based [10], and neural-network-based [16] approach were able to cross the 80% threshold of selected metric with 2-class or 3-class classification. 5-class classification achieved approximately 40% in selected metrics. Successful 5-class (or finer)

---

[1] https://github.com/okruhlica/SlovakSentimentLexicon.

classification is yet to be achieved, multi-class classification with lexicon-based methods is unexplored. There are not many freely accessible Slovak sentiment lexicons. We were able to obtain Ukruhlica's lexicon [14], and Slovak lexicon which was the result of a larger effort by Chen and Skiena [6] (SLf81L). Collecting sentiment annotations using crowdsourcing is not unheard of [13], so some success may be expected with our GWAP.

3   Game for Sentiment Collection

We created a GWAP to collect sentiment annotations from human annotators. *Sentižrút*[2] (the name of our game) is intended for a single player, who is supposed to annotate the sentiment of words in the current sentence one after another. This means that the further a player proceeds within a sentence, the more context they have to consider when assigning sentiment to the current word. Players are not explicitly instructed whether to consider the context of previously seen words or not. This means that the annotations we collect may come from unrestricted human perception. This might be advantage over collecting sentiment annotations using best-worst scaling [9], although there is no guarantee of increased data quality. After all words of a sentence are annotated, the player is additionally prompted to select the sentiment of the sentence as a whole. Both sentiments (the sentiment of a word and the sentiment of a whole sentence) are categorical values on a 5-point scale, numerically represented as $-2, -1, 0, 1,$ 2 respectively. Core game loop of our game is shown as an activity diagram in Fig. 1.

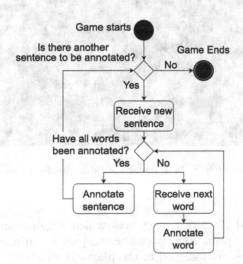

Fig. 1. Core game loop of Sentižrút.

---

[2] The name could be loosely translated to *Sentiment Devourer*.

By assigning sentiment to a word or a sentence, the player earns points. Screen for assigning sentiment is illustrated in Fig. 2. Gathering a sufficient amount of points allows the player to *level up* and obtain a new skill in the skill tree (Fig. 3). Skills increase income of points earned by annotating. The skill tree offers different branches with specific philosophies. The game also features achievements which grant a one-time bonus in game points. Players are ranked in a global leaderboard, with notifications indicating ranking changes.

Fig. 2. Basic playing screen of *Sentižrút*. Player chooses sentiment of word from illustrated scale. Left panel contains notifications, in this case about improved leaderboard ranking.

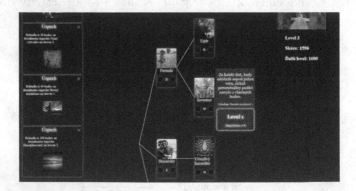

Fig. 3. Skill tree of *Sentižrút*. Left panel contains notifications, in this case about reached achievements.

The game offered 170 Slovak sentences which were manually collected from various web sources, primarily reviews. Each player can annotate all the sentences. The sentences are assigned to the player in random order.

Within a week, 201 players played the game while 126 of them annotated at least one sentence. Top 10 players contributed with 74% of collected annotations. This includes one of the researchers, to reinforce quality of annotations, similarly to what was done in [7]. Most players were probably attracted by intrinsic motivation [22], as many of them were university students aware that they are contributing to research. The leaderboard served as a boost to extrinsic motivation [22]. Offering money for participation could have increased volume of obtained data [22], but the game-like feeling would likely be suppressed.

4   Experiments and Evaluation

After lemmatization using an online NLP tool for Slovak[3], we had 920 distinct words in the dataset for which we collected annotations. We collected a total of 12 593 word annotations and 1 132 sentence annotations. Figure 4 demonstrates distribution of count of acquired annotations per player on logarithmic scale (of base 2). Players without any contribution are omitted.

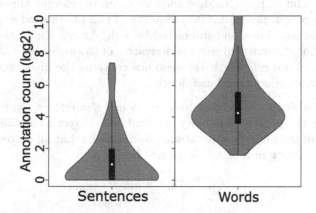

Fig. 4. Acquired annotation count per player on logarithmic scale (violin plot).

Table 1 shows that sentiment categories with higher intensity are less numerous for both sentences and words, with one exception. *Neutral* category is significantly more numerous than other categories for words, as expected.

Table 1. Annotation count of both words and sentences per categories.

Category	−2	−1	0	1	2	Total
Sentence	48	224	342	401	117	**1 132**
Word	222	742	9 491	1 653	485	**12 593**

---

[3] http://arl6.library.sk/nlp4sk/.

4.1 Sentiment Lexicon

The acquired data was used to construct a sentiment lexicon. Annotation value v of each word is defined by Eq. 1.

$$v = \frac{w_1 * v_{word} + w_2 * v_{sentence}}{w_1 + w_2} \tag{1}$$

There, v_{word} is value derived from word annotations, $v_{sentence}$ is value derived from sentence annotations, and w_1, w_2 are their corresponding weights. Value derived from annotations on given level (word or sentence) is defined by Eq. 2.

$$v_{level} = \frac{\sum_{i=1}^{N} \sum_{j=1}^{M_i} pr_i * av_{ij}}{\sum_{i=1}^{N} M_i * pr_i} \tag{2}$$

N stands for count of players, M_i stands for count of relevant annotations submitted by i-th player, pr_i stands for reliability of i-th player, and av_{ij} stands for j-th relevant annotation value submitted by i-th player. Annotations on word level are relevant if annotated word is derivative of the given word. Annotations on sentence level are relevant if the sentence contains the given word. Lexicon construction was parametrized as following:

- **player reliability** - possible values are: *None*, *Discrete* or *Continuous*. This defines how the player's reliability is calculated. Player's reliability serves as the weight of the player's annotations. *None* means that all players are equal. *Discrete* reliability of a player is defined by Eq. 3.

$$r = \frac{\sum_{i=1}^{N} \begin{cases} 1, & \text{if } av_i = av_{i\_avg} \\ 0, & \text{otherwise} \end{cases}}{N} \tag{3}$$

N is count of annotations submitted by the player, av_i is value of player's i-th annotation, and av_{i_avg} is average annotation value of given word or sentence. *Continuous* reliability of a player is defined by Eq. 4. Note that 5 is the maximal possible annotation value.

$$r = \frac{\sum_{i=1}^{N}(1 - \frac{|av_i - av_{i\_avg}|}{5})}{N} \tag{4}$$

- **word annotation weight** - value from interval *(0; 1)*. Category of a word is calculated as a weighted average of value derived from word annotations and value derived from sentence annotations. This value defines the weight of value derived from word annotations w_{word}, $w_{sentence} = 1 - w_{word}$ is then the weight of value derived from sentence annotations.

Sentiment of a sentence is determined from sentiment of its words, differently for each method. The first method defines sentiment of a sentence s_s by Eq. 5.

$$s_s = \frac{\sum_{i=1}^{N} s_{wi}}{N} \tag{5}$$

N is count of words in sentence, and s_{wi} is sentiment value of i-th word taken from lexicon. The second method is similar, but ignores neutral sentiment. If there is no sentiment available, the sentence is marked as neutral. The third method defines sentiment of a sentence s_s by Eq. 6.

$$s_s = \begin{cases} -L, & \text{if } -L > \sum_{i=1}^{N} s_{wi} \\ L, & \text{if } L < \sum_{i=1}^{N} s_{wi} \\ \sum_{i=1}^{N} s_{wi}, & \text{otherwise,} \end{cases} \qquad (6)$$

N is count of words in sentence, s_{wi} is sentiment value of i-th word taken from lexicon, and L is limit value (1 with 3-class classification, 2 with 5-class classification, etc.). Negators, intensifiers nor diminishers are not considered. Note that our aim is not to design a perfect sentiment analysis method, but to compare the usefulness of lexicons.

We compared results of our best lexicon with Okruhlica's lexicon [14] (5897 words[4]) and SLf81L [6] (2428 words). Overlap of the lexicons is illustrated in Fig. 5.

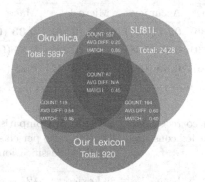

Fig. 5. Lexicon overlap. *COUNT* stands for count of overlapping words, *AVG DIFF* stands of MAE between overlapping word sentiment and *MATCH* stands for ratio of overlapping words that have same sentiment. Note that the displayed overlap of all three lexicons and overlaps of pairs of lexicons are not disjoint.

Evaluation was performed towards a dataset of mobile phone reviews, which we manually annotated for this purpose. The dataset consists of 1220 sentences and does not overlap with the sentences annotated by player. The results of 3-class classification are displayed in Table 2 and results of 5-class classification are shown in Table 3. Each value in a table is average of the three sentiment analysis methods. Working with average of several methods allowed us to evaluate the lexicons more generally, as sticking to a single method would necessarily fit the data to specifics of the method. The values in brackets represent results on dataset cleared from sentences where at least one of the lexicons does not contain any words of the sentence.

---

[4] Over 100 words with intensity 3 were removed.

46   L. Radoský and M. Blšták

Table 2. Comparison of metrics of lexicons used for 3-class classification.

Lexicon	Precision	Accuracy	Micro F1	Macro F1	Recall
Okruhlica	**0.57** (0.42)	0.58 (0.47)	0.58 (0.47)	**0.48** (0.39)	**0.48** (0.43)
SLf81L	0.53 (**0.45**)	0.52 (**0.48**)	0.52 (**0.48**)	0.47 (**0.42**)	**0.48** (**0.44**)
Our Lex	0.49 (0.41)	**0.59** (0.47)	**0.59** (0.47)	0.39 (0.36)	0.42 (0.42)

While values are lower for 5-class classification, relationships between lexicons are the same in both cases. Okruhlica's lexicon [14] outperforms SLf81L [6] in all metrics. Both lexicons outperform our lexicon in precision, macro F1, and recall. Our lexicon has the best accuracy and micro F1, although the difference with Okruhlica's lexicon [14] is small. Overall, we consider Okruhlica's lexicon the best, however our lexicon is not far behind and surpasses it in some metrics.

Table 3. Comparison of metrics of lexicons used for 5-class classification.

Lexicon	Precision	Accuracy	Micro F1	Macro F1	Recall
Okruhlica	**0.54** (0.36)	0.55 (0.36)	0.55 (0.36)	**0.33** (**0.27**)	**0.34** (**0.32**)
SLf81L	0.50 (**0.39**)	0.47 (0.38)	0.47 (0.38)	0.27 (0.23)	0.28 (0.25)
Our Lex	0.47 (0.36)	**0.58** (**0.45**)	**0.58** (**0.45**)	0.23 (0.22)	0.25 (0.26)

Table 4. Comparison of correct classification ratio per class of lexicons used for 3-class classification.

Lexicon	−1	0	1
Okruhlica	0.18	0.73	0.54
SLf81L	**0.29**	0.55	**0.59**
Our Lex.	0.11	**0.91**	0.23
Okruhlica	**0.38**	0.09	**0.84**
SLf81L	0.32	0.22	0.77
Our Lex.	0.11	**0.85**	0.30

Table 5. Comparison of correct classification ratio per class of lexicons used for 5-class classification.

Lexicon	−2	−1	0	1	2
Okruhlica	**0.06**	0.17	0.73	0.40	**0.39**
SLf81L	0	**0.27**	0.55	**0.49**	0.11
Our Lex.	0	0.11	**0.91**	0.21	0.03
Okruhlica	**0.07**	**0.36**	0.09	**0.63**	**0.49**
SLf81L	0	0.28	0.22	**0.63**	0.14
Our Lex.	0	0.11	**0.85**	0.27	0.04

For all lexicons, we also consider correct classification ratio per class, defined by Eq. 7.

$$cr_c = \frac{\sum_{i=1}^{3} \frac{|\hat{Y}_{ci}|}{|Y_{ci}|}}{3} \tag{7}$$

$\hat{Y}_{ci}$ stands for set of correct predictions by i-th method for samples of class c and Y_{ci} is set of all samples of class c present in dataset. The results of 3-class classification are shown in Table 4 and results of 5-class classification are shown in Table 5 - values above line. In both cases, SLf81L has the best score for classes with intensity 1, our lexicon has the best score with neutral class. Okruhlica's lexicon dominates in classification of classes with intensity 2 (only 5-class classification). Results are weaker for negative sentences than for positive sentences.

Sentence with no available sentiment from lexicon is considered neutral. Our lexicon's high score with neutral sentences might therefore indicate that it simply had no words available in many cases. However, the truth is quite the opposite. As Table 6 shows, our lexicon has by far the lowest ratio of sentences where it did not offer sentiment for any of the words (full miss ratio) and the highest ratio of words in the sentence for which the lexicon offered sentiment (hit ratio). This is to be expected, as we collected annotations for entire sentences, while Okruhlica focused on opinion words, i.e. adjectives and proverbs. This helped our lexicon to achieve great results with neutral sentences. In other sentences, too much information about neutral words hindered its results.

Table 6. Dataset vocabulary coverage by lexicons.

Lexicon	Full miss ratio	Average hit ratio
Okruhlica	58%	5%
SLf81L	35%	9%
Our Lex.	**2%**	**48%**

This is why we also evaluated lexicons on a dataset version where every lexicon contains at least one word of each sentence. This dataset contained 444 sentences. As Table 2 shows (values in brackets), SLf81L achieved score superior to other lexicons in all metrics with 3-class classification. This is because most neutral sentences were removed, so both our and Okruhlica's lexicon lost advantage due to more balanced class representation. As Table 3 shows (values in brackets), SLf81L achieved superior precision in 5-class classification, but lexicon order is otherwise the same as with full dataset. This is mostly due to SLf81L's poor ability to identify classes with intensity 2, as it does not contain any words with that intensity. When looking at individual classes (Tables 4 and 5 - values below line), SLf81L looses its position to Okruhlica's lexicon with classes with intensity 1. Our lexicon's superiority in neutral sentence identification is magnified.

5 Conclusion

In this paper, we have described the process of collecting sentiment analysis annotations using crowdsourcing approach via a game with a purpose. We were able to obtain over 12 000 word annotations and over 1 000 sentence annotations within a week. We utilized artificial bee colony algorithm to construct optimal sentiment lexicon from the collected annotations. To prove its quality, three simple sentence-level sentiment analysis methods using this lexicon were applied on a manually annotated dataset. The same evaluation method was applied to other sentiment lexicons. Results show that compared to other lexicons, our lexicon achieves weaker scores in some metrics, but better accuracy and micro F1 scores. Our lexicon has been proven to be especially useful when identifying neutral sentences. It is noteworthy that being the smallest one did not impair our lexicon's usefulness compared to other lexicons. The annotated evaluation dataset and constructed lexicon were made available for future researchers[5].

It would be interesting to collect more data from annotators, providing them wider selection of sentences using wider vocabulary. The activity might serve as CAPTCHA, potentially allowing to collect data at large scale, for example under hCaptcha[6]. Results of our work might be used to propose a sentiment analysis method using all analyzed lexicons, taking advantage of their different properties.

References

1. von Ahn, L., Dabbish, L.: Labeling images with a computer game. In: Proceedings of the SIGCHI Conference on Human Factors in Computing Systems (CHI '04), pp. 319–326. Association for Computing Machinery, New York (2004)
2. Alqaryouti, O., Siyam, N., Shaalan, K.: A sentiment analysis lexical resource and dataset for government smart apps domain. In: Hassanien, A.E., Tolba, M.F., Shaalan, K., Azar, A.T. (eds.) AISI 2018. AISC, vol. 845, pp. 230–240. Springer, Cham (2019). https://doi.org/10.1007/978-3-319-99010-1_21
3. Angwin, J., Larson, J., Mattu, S., Kirchner, L.: Machine bias, May 2016. https://www.propublica.org/article/machine-bias-risk-assessments-in-criminal-sentencing. Accessed 23 May 2016
4. Bouazizi, M., Ohtsuki, T.: Sentiment analysis: from binary to multi-class classification: a pattern-based approach for multi-class sentiment analysis in twitter. In: 2016 IEEE International Conference on Communications (ICC), pp. 1–6 (2016)
5. Bučar, J., Žnidaršič, M., Povh, J.: Annotated news corpora and a lexicon for sentiment analysis in Slovene. Lang. Resour. Eval. **52**, 895–919 (2018)
6. Chen, Y., Skiena, S.: Building sentiment lexicons for all major languages. In: 52nd Annual Meeting of the Association for Computational Linguistics, ACL 2014 - Proceedings of the Conference, vol. 2, pp. 383–389, June 2014
7. Duwairi, R.M.: Arabic sentiment analysis using supervised classification. In: Proceedings - 2014 International Conference on Future Internet of Things and Cloud, FiCloud 2014, August 2014
8. Hatzivassiloglou, V., McKeown, K.R.: Predicting the semantic orientation of adjectives. In: Proceedings of the 35th Annual Meeting of the Association for Computational Linguistics and Eighth Conference of the European Chapter of the Association for Computational Linguistics, ACL 1998/EACL 1998, pp. 174–181. Association for Computational Linguistics, USA (1997)

---

[5] https://github.com/Neltharion59/SlovakSentimentLexicon.
[6] https://www.hcaptcha.com/.

9. Kiritchenko, S., Mohammad, S.M.: Capturing reliable fine-grained sentiment associations by crowdsourcing and best-worst scaling. In: Proceedings of the 2016 Conference of the North American Chapter of the Association for Computational Linguistics: Human Language Technologies, pp. 811–817. Association for Computational Linguistics, San Diego, California, June 2016

10. Krchnavy, R., Simko, M.: Sentiment analysis of social network posts in Slovak language. In: 2017 12th International Workshop on Semantic and Social Media Adaptation and Personalization (SMAP), pp. 20–25 (2017)

11. Liu, B.: Sentiment Analysis: Mining Opinions, Sentiments, and Emotions. Cambridge University Press, Cambridge (2015)

12. Mikula, M., Machova, K.: Combined approach for sentiment analysis in Slovak using a dictionary annotated by particle swarm optimization. Acta Electrotech. Inf. **18**, 27–34 (2018)

13. Musat, C.C., Ghasemi, A., Faltings, B.: Sentiment analysis using a novel human computation game. In: Proceedings of the 3rd Workshop on the People's Web Meets NLP: Collaboratively Constructed Semantic Resources and their Applications to NLP, pp. 1–9. Association for Computational Linguistics, Jeju, July 2012. https://aclanthology.org/W12-4001

14. Okruhlica, A.: Slovak sentiment lexicon induction in absence of labeled data. Master's thesis, Comenius University Bratislava (2013)

15. Pang, B., Lee, L.: Opinion mining and sentiment analysis. Found. Trends Inf. Retr. **2**(1–2), 1–135 (2008)

16. Pecar, S., Simko, M., Bielikova, M.: Improving sentiment classification in Slovak language. In: Proceedings of the 7th Workshop on Balto-Slavic Natural Language Processing, pp. 114–119. Association for Computational Linguistics, Florence, August 2019

17. Rudin, C.: Stop explaining black box machine learning models for high stakes decisions and use interpretable models instead. Nat. Mach. Intell. **1**, 206–215 (2019)

18. Simko, J., Bieliková, M.: Semantic Acquisition Games - Harnessing Manpower for Creating Semantics. Springer, Heidelberg (2014). https://doi.org/10.1007/978-3-319-06115-3

19. Speer, R.: ConceptNet numberbatch 17.04: better, less-stereotyped word vectors, April 2017. http://blog.conceptnet.io/posts/2017/conceptnet-numberbatch-17-04-better-less-stereotyped-word-vectors/. Accessed 24 Apr 2017

20. Xu, L., Chamberlain, J.: Cipher: a prototype game-with-a-purpose for detecting errors in text. In: Workshop on Games and Natural Language Processing, pp. 17–25. European Language Resources Association, Marseille, May 2020

21. You, Q.: Sentiment and emotion analysis for social multimedia: methodologies and applications. In: Proceedings of the 24th ACM International Conference on Multimedia, MM 2016, pp. 1445–1449. Association for Computing Machinery, New York (2016)

22. Zichermann, G., Linder, J.: Game-Based Marketing: Inspire Customer Loyalty Through Rewards, Challenges, and Contests. Wiley, Hoboken (2010)

23. Šimko, J., Tvarožek, M., Bieliková, M.: Human computation: image metadata acquisition based on a single-player annotation game. Int. J. Hum. Comput. Stud. **71**(10), 933–945 (2013)

Robustness of Named Entity Recognition: Case of Latvian

Rinalds Viksna[1,2(✉)] and Inguna Skadiņa[1,2]

[1] Tilde, Vienibas gatve 75a, Riga 1004, Latvia
{rinalds.viksna,inguna.skadina}@tilde.lv
[2] Faculty of Computing, University of Latvia, Riga, Latvia

Abstract. Recent developments in Named Entity Recognition (NER) have demonstrated good results for grammatically correct texts, even in low resourced settings. However, when the NER model faces ungrammatical text, it often shows poor performance. In this study, we analyze NER performance on datasets containing errors typical for user-generated texts in the Latvian language. We explore three different strategies to increase the robustness of the named entity recognition: error injection into grammatically correct texts, augmenting grammatically correct texts with erroneous texts and augmenting grammatically correct texts with erroneous texts that contain specific types of errors. We demonstrate that in low resourced settings, the best noise-robust model could be obtained by augmenting training data with datasets containing different error types. Our best model achieves an average F1 score of 83.5 (84.1 for baseline) on grammatically correct text, while keeping good performance (79 F1 vs. 66 for baseline) on noisy texts.

Keywords: Named entity recognition · NER · Robustness · Low resourced settings · Ungrammatical text

1 Introduction

With the introduction of BERT models [8], substantial progress has been made on the task of named entity recognition (NER), not only for widely spoken languages, but also for less resources languages, e.g., Finnish [22], Portuguese [19] or Slavic [1] languages. In the case of the Latvian language, BERT models [21, 24] achieve F1 scores of over 82 when compared to the previous research [16, 23] with F1 scores of 60–78.

The reported results have been obtained on grammatically correct datasets. However, in real life user-generated content (e.g., Twitter records, conversations in chat platforms, etc.) or automatically generated text (e.g., optical character recognition (OCR), output from the speech recognizer or machine translation system) usually contains multiple errors.

The issue of noise in output of automatic text processing tools has been recognized for many years: the output from speech recognizers usually lacks capitalization and punctuation and could even be ungrammatical [5, 20], while text obtained using OCR could also contain character errors [11]. The global popularity of mobile devices, social media and

© Springer Nature Switzerland AG 2021
L. Espinosa-Anke et al. (Eds.): SLSP 2021, LNAI 13062, pp. 50–58, 2021.
https://doi.org/10.1007/978-3-030-89579-2_5

virtual assistants lead to significant amount of user generated text, which is being actively explored in different natural processing tasks. However, informal, user-generated text often suffers from inconsistent capitalization, usage of abbreviations typical for informal conversations [18], non-standard or misspelled words, and repeated letters [2].

Named entity recognizers trained on literary texts lose their quality dramatically when applied to ungrammatical or non-literal texts. For instance, Mayhew et al. [12] reported a performance drop from an F1 score of 92 to an F1 score of 34, when the cased NER model is applied to uncased text.

Possible solutions to noisy casing include the use of a truecaser [13], augmenting data with a cased and uncased version of the data [12] and lowercasing both training and test data [4]. Possible solutions to spelling errors and non-standard words include text normalization [2, 7], augmentation of training data with various types of common user errors [3] and adding noise to the models [14]. In case of resource rich languages, e.g., English, domain specific pre-trained language models [15] and domain adaptive pretraining [10] have demonstrated state of the art results.

In this paper, we explore three data augmentation methods to build an NER model in low resourced settings which is robust to both - casing and orthographical errors. We demonstrate that the augmentation of training data with various errors allows one to build a single model that performs well on both - well-formed and erroneous texts.

2 Data

For our experiments we use the Balanced State-of-the-Art Multilayer Corpus for NLU [9]. The corpus contains many different grammatically correct texts of various types: around 60% come from news sources, around 20% is fiction, around 10% are legal texts, around 5% is spoken language (transcripts), and the rest is miscellaneous. The total size of the corpus is 3947 paragraphs or 11425 sentences. It includes named entity annotations of 9697 named entities (3,104 person, 2,031 GPE, 1,847 organization, 1,227 time, 677 location, 293 product, 259 event, 215 entity, 44 money). This 9-entity tagset was converted to a 4-entity tagset in accordance with MUC-7 [6] entity names subtask, keeping PERSON and ORGANIZATION, joining LOCATION and GPE into LOCATION and joining the rest of the classes into the MISC category. The obtained corpus was then split into training (7807 entities) and test (1890 entities) sets, as shown in Table 1.

Table 1. Named entity distribution in training and test datasets.

NE category	Training data	Test data
PERSON	2535	569
ORGANIZATION	1481	366
LOCATION	2157	551
MISC	1634	404
Total	**7807**	**1890**

Since a manually labeled dataset, containing various error types is not available for Latvian, we introduce artificial errors into corpus.

Table 2. Error types, illustration for sentence "Šis (this) teikums (sentence) satur (contain) daudz (many), daudz (many) kļūdu (errors)".

Error type	Explanation	Example
Remove punct	Omit punctuation	Šis teikums satur daudz_ daudz kļūdu_
Add comma	After random word add a comma if there is no punctuation already with probability P = 0.1	Šis, teikums satur daudz, daudz kļūdu
diacr aa zh	Latinize all diacritics phonetically	Shis teikums satur daudz, daudz kljuudu
diacr a z	Latinize all diacritics using the same letter without diacritic	Sis teikums satur daudz, daudz kludu
Double letter	Repeat a letter with probability P = 0.05	Šiss teiikums satur dauddz, daudz kļļūdu
Insert letter	After every letter insert another random letter with probability P = 0.05	Šis tefikums satur damudz, daudz Ukļūdu
Omit letter	Delete a letter with probability P = 0.05	Šis tei_ums s_tur daudz, da_dz kļūdu
Substitute letter	Substitute a letter with another random letter with probability P = 0.05	Šis teikuma satur daydz, daudz jļūdu
Swap letters	Swap two adjacent letters with probability P = 0.1	Šis tiekums satru daudz, adudz kļūdu
Uppercase random	Uppercase a letter with probability P = 0.1	Šis teIkums Satur dAudz, daudz kĻūdu
Switch case	Switch case of a letter with probability P = 0.1	šiS teikums satur dAudz, daudz kļūdu
Uppercase all	Uppercase all text	ŠIS TEIKUMS SATUR DAUDZ, DAUDZ KĻŪDU
Capitalize case	Capitalize every word	Šis Teikums Satur Daudz, Daudz Kļūdu
Lower case	Lowercase all text	šis teikums satur daudz, daudz kļūdu

We follow error classification in user generated texts introduced by Deksne [7]. This classification is based on analysis of the Latvian Tweet corpus [17]. The following error types were identified in this corpus: dropped vowels, switched letters, missing diacritics, words written together, shortenings and slang words and various capitalization and typographic errors. In addition to these error types, we create capitalized, lowercased, uppercased, and randomly cased versions of text. In total, 14 classes of errors are introduced, as illustrated in Table 2. Each error type is introduced into our training and test datasets.

3 Models

In our experiments we create Latvian NER models with dense and Conditional Random Field (CRF) layers on top of pre-trained cased Latvian BERT (12 layers, 12 attention heads, hidden size 768) [21]. NER models are finetuned for 12 epochs, using the sequence length of 128, train batch size of 16, and learning rate of 2e−5.

3.1 Impact of Error Types on NER

To assess the impact of each error type on the model's performance, we trained 15 NER models: baseline model on original data and 14 models for each error type on the dataset with the respective error. Evaluation results for these NER models are summarized in Table 3.

Our evaluation results show that some error types have a significant impact on the model's performance, while some error types have surprisingly little impact on the model's predictive capabilities. In particular:

- Model trained on original data demonstrates a dramatic performance drop when applied to lowercased, uppercased, or capitalized text.
- Random letter insertions, deletions, substitutions, and permutations reduce the model's performance by 5–10 F1 points. Letter permutations in pronouns could create a new word, which could be a person name, or the named entity is so distorted that it is no longer recognizable.
- In the Latvian language, diacritics are important (e.g. "mašīna" (car) vs. "māsiņa" ("nurse" and also diminutive of "sister")). When all diacritics are replaced with doubled letters, performance drops by 17 F1 points, but, when all diacritics are replaced by single letters, performance drops by 11 F1 points.
- NER model is quite robust to punctuation errors - randomly added commas only reduce the performance of the model by 2 F1 points, while the complete absence of punctuation leads to a drop of 4 F1 points.

Table 3. Performance of NER models (F1) trained on data with errors. Rows: NER models trained on datasets with added noise. Columns: Test sets with added noise.

	original data	remove punct	add comma	diacr aa zh	diacr a z	double letter	insert letter	omit letter	substitute letter	swap letters	uppercase random	switch case	uppercase all	capitalize case	lower case
baseline	84	80	82	67	74	78	77	76	74	66	64	58	36	41	32
remove punct	72	83	70	56	62	67	64	62	63	55	52	48	26	28	23
add comma	84	80	83	68	75	78	76	74	74	66	63	57	35	35	31
diacr aa zh	82	76	79	81	78	79	78	75	75	69	61	56	34	26	31
diacr a z	83	77	80	76	80	78	77	75	75	69	61	56	36	30	30
double letter	84	79	81	74	76	81	80	78	78	71	65	59	39	35	32
insert letter	84	79	81	74	76	80	80	77	78	72	64	59	40	30	30
omit letter	83	79	81	73	76	80	79	79	77	73	64	61	39	31	52
substitute letter	84	79	81	73	76	81	80	78	79	73	64	60	40	32	53
swap letters	84	79	81	74	77	80	80	79	77	76	52	48	39	27	32
uppercase random	82	78	81	72	76	80	79	77	76	71	77	71	44	67	40
switch case	82	77	80	66	74	79	77	77	75	69	77	74	44	67	69
uppercase all	42	35	42	38	47	47	45	46	44	42	50	49	66	48	40
capitalize case	53	51	52	40	54	51	48	50	48	43	50	46	38	78	44
lower case	62	57	60	26	43	48	42	47	41	32	35	35	35	39	80

Most of the models trained with erroneous data perform well on corresponding test sets and the original test set, however, models trained on data with broken casing or no punctuation perform poorly on original and most other datasets. These four datasets are the ones where an important feature (casing, punctuation) is removed completely. Our model is even weaker on uppercased data when compared to lowercased data. One reason could be the influence of the BERT model, which has been trained on mostly orthographically correct texts, which are mostly lowercased. On the other hand, the case marker is a very important feature for named entity recognition in Latvian. For instance, "es" (I) is a pronoun in Latvian, while "ES" (EU – European Union) is an abbreviation.

3.2 Data Augmentation and Robustness

Next, we use original training data and merge it with datasets for each type of errors and train and test corresponding models. The results of this experiment are consolidated in Table 4. Our models demonstrate good results on the original test set and respective erroneous test set. However, for some error types (letter permutations, diacritics replacement, comma addition) we obtained similar results to the previous experiment with only a little improvement.

Table 4. Performance of NER models (F1) trained on a combination of baseline data and noisy data. Rows: NER models finetuned on datasets augmented with noise. Columns: Test sets with added noise.

	original data	remove punct	add comma	diacr aa zh	diacr a z	double letter	insert letter	omit letter	substitute letter	swap letters	uppercase rand	switch case	uppercase all	capitalize case	lower case
Baseline	84	81	82	69	74	78	78	74	75	68	62	60	27	40	35
remove punct	84	83	82	67	73	76	76	74	73	67	59	58	28	35	33
add comma	84	81	85	67	73	77	77	74	73	67	61	59	33	38	35
diacr aa zh	84	81	81	81	79	78	80	76	76	73	63	60	38	37	28
diacr a z	85	80	82	77	80	79	79	76	77	71	62	60	37	33	30
double letter	84	80	82	74	77	82	80	77	78	73	62	60	36	28	29
insert letter	84	79	82	75	77	81	80	77	78	73	63	61	39	35	31
omit letter	84	81	82	72	78	80	80	79	77	73	63	61	34	29	48
substitute letter	84	79	82	73	77	80	80	78	78	74	60	60	40	34	50
swap letters	84	79	82	75	77	81	80	78	78	77	53	52	38	31	33
uppercase random	84	80	82	73	77	81	80	77	78	72	76	73	44	66	36
switch case	85	80	82	68	75	79	78	78	78	73	77	75	45	67	65
uppercase all	84	80	81	68	75	79	77	75	75	71	68	66	69	57	44
capitalize case	84	80	81	62	73	75	75	74	72	66	67	66	36	79	51
lower case	84	79	81	33	55	62	56	61	55	40	47	48	41	56	81

This experiment allows us to conclude, that for all types of errors analyzed in this paper it is possible to augment the generic dataset in a way that allows the training of a model which is resistant to the specific type of errors, while performing well on original, orthographically correct text.

3.3 Finding a Robust Model

Finally, we finetune our model using three datasets (see Table 5). Model1 is trained on a dataset obtained by merging training data with all erroneous training datasets. Model2 is trained on all datasets which demonstrated low performance in the first experiment (datasets with phonetically replaced diacritics, no punctuation, uppercased, lowercased data) and original data. Model3 is trained using an original dataset together with uppercased, lowercased, capitalized datasets and both datasets with replaced diacritics, doubled letters and removed punctuation.

Table 5. Datasets used for final robust model training

	Clean data	Remove punctuation	Add comma	Diacritics (aa, zh)	Diatritics (a,z)	Double letter	Insert letter	Omit letter	Substitute letter	Swap letters	Uppercase random	Switch case	Uppercase all	Capitalize all	Capitalize case	Lower case
Model1	+	+	+	+	+	+	+	+	+	+	+	+	+	+	+	+
Model2	+	+		+									+			+
Model3	+	+		+	+	+							+	+		+

All three models were evaluated on all test datasets, the results are provided in Fig. 1. We can see that the original baseline system performs worst overall, with an averaged F1 score of 66,02. However, it is best when used on original, grammatically correct data reaching an F1 score of 84.12. On the other hand, all four models work well on an original dataset, with the worst, Model3, achieving an 82.78 F1 score. The best, Model1, achieves an F1 score of 79,47 on average and is almost as good on the original dataset as the baseline, reaching 83.50 F1 score for grammatically correct data.

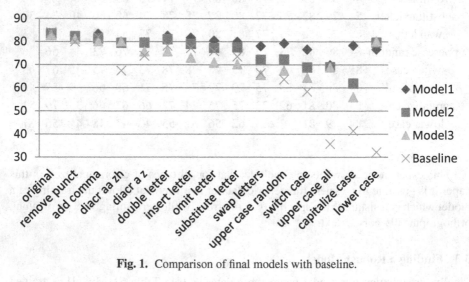

Fig. 1. Comparison of final models with baseline.

4   Conclusions

We have performed a systematic analysis of data augmentation methods with the goal of improving the performance and robustness of the Latvian NER. We demonstrated that the most effective strategy is to augment training data with adversarially generated noisy data. The resulting model has almost as good performance on original data, while being significantly better on noisy and erroneous data.

Results presented in this paper were obtained on artificially created noisy data. In the future we would like to test our approach on real user generated content, e.g., Twitter or dialog system data. Another direction of future work is to test hypothesis that the proposed approach could be extended to other less resourced, highly inflected languages.

The data used in this work and code used to insert errors is available on GitHub[1].

Acknowledgements. This research has been supported by the European Regional Development Fund within the joint research project of SIA Tilde and University of Latvia "Multilingual Artificial Intelligence Based Human Computer Interaction" No. 1.1.1.1/18/A/148.

References

1. Arkhipov, M., Trofimova, M., Kuratov, Y., Sorokin, A.: Tuning multilingual transformers for language-specific named entity recognition. In: Proceedings of the 7th Workshop on Balto-Slavic Natural Language Processing, pp. 89–93 (2019). http://dx.doi.org/10.18653/v1/W19-3712
2. Baldwin, T., de Marneffe, M.C., Han, B., Kim, Y.-B., Ritter, A., Xu, W.: Shared tasks of the 2015 workshop on noisy user-generated text: Twitter lexical normalization and named entity recognition. In: Proceedings of the Workshop on Noisy User-generated Text, pp. 126–135 (2015). http://dx.doi.org/10.18653/v1/W15-4319
3. Bergmanis, T., Stafanovičs, A., Pinnis, M.: Robust neural machine translation: modeling orthographic and interpunctual variation. In: Human Language Technologies – The Baltic Perspective, pp. 80–86 (2020). https://doi.org/10.3233/FAIA200606
4. Bodapati, S., Yun, H., Al-Onaizan, Y.: Robustness to capitalization errors in named entity recognition. In: Proceedings of the 2019 EMNLP Workshop W-NUT: The 5th Workshop on Noisy User-generated Text, pp. 237–242 (2019). http://dx.doi.org/10.18653/v1/D19-5531
5. Brown, E.W., Coden, A.R.: Capitalization recovery for text. In: Coden, A.R., Brown, E.W., Srinivasan, S. (eds.) IRTSA 2001. LNCS, vol. 2273, pp. 11–22. Springer, Heidelberg (2002). https://doi.org/10.1007/3-540-45637-6_2
6. Chinchor, N.: MUC-7 named entity task definition. In: Seventh Message Understanding Conference (MUC-7): Proceedings of a Conference Held in Fairfax, Virginia, 29 April–1 May 1998 (1998). https://www.aclweb.org/anthology/M98-1028.
7. Deksne, D.: Chat language normalisation using machine learning methods. In: Proceedings of the 11th International Conference on Agents and Artificial Intelligence (ICAART 2019), pp. 965–972 (2019). https://doi.org/10.5220/0007693509650972
8. Devlin, J., Chang, M.-W., Lee, K., Toutanova, K.: BERT: pre-training of deep bidirectional transformers for language understanding. In Proceedings of the 2019 Conference of the North American Chapter of the Association for Computational Linguistics: Human Language Technologies, Volume 1 (Long and Short Papers), pp. 4171–4186 (2019). https://www.aclweb.org/anthology/N19-1423/
9. Gruzitis, N., et al.: Creation of a balanced state-of-the-art multilayer corpus for NLU. In: Proceedings of the Eleventh International Conference on Language Resources and Evaluation (LREC 2018) (2018). https://www.aclweb.org/anthology/L18-1714

---

[1] https://github.com/RinaldsViksna/NER-data-augmentation.

10. Gururangan, S., et al.: Don't stop pretraining: adapt language models to domains and tasks. In: Proceedings of the 58th Annual Meeting of the Association for Computational Linguistics, pp. 8342–8360 (2020). https://aclanthology.org/2020.acl-main.740.pdf

11. Hamdi, A., Jean-Caurant, A., Sidere, N., Coustaty, M., Doucet, A.: An analysis of the performance of named entity recognition over OCRed documents. In: ACM/IEEE Joint Conference on Digital Libraries (JCDL) (2019). https://doi.org/10.1109/JCDL.2019.00057

12. Mayhew, S., Tsygankova, T., Roth, D.: ner and pos when nothing is capitalized. In: Proceedings of the 2019 Conference on Empirical Methods in Natural Language Processing and the 9th International Joint Conference on Natural Language Processing (EMNLP-IJCNLP), pp. 6256–6261 (2019). http://dx.doi.org/10.18653/v1/D19-1650

13. Mayhew, S., Gupta, N., Roth, D.: Robust named entity recognition with truecasing pretraining. In: Proceedings of the AAAI Conference on Artificial Intelligence, pp. 8480–8487 (2020). https://doi.org/10.1609/aaai.v34i05.6368

14. Narayan, P.L., Nagesh, A., Surdeanu, M.: Exploration of noise strategies in semi-supervised named entity classification. In: Proceedings of the Eighth Joint Conference on Lexical and Computational Semantics (*SEM), pp. 186–191 (2019). http://dx.doi.org/10.18653/v1/S19-1020

15. Nguyen, D.Q., Vu, T., Nguyen, A.T.: BERTweet: a pre-trained language model for English Tweets. In: Proceedings of the 2020 Conference on Empirical Methods in Natural Language Processing: System Demonstrations, pp. 9–14 (2020). https://aclanthology.org/2020.emnlp-demos.2/

16. Pinnis, M.: Latvian and Lithuanian named entity recognition with TildeNER. In: Proceedings of the Eighth International Conference on Language Resources and Evaluation (LREC 2012), pp. 1258–1265 (2012). https://www.aclweb.org/anthology/L12-1566/

17. Pinnis, M.: Latvian tweet corpus and investigation of sentiment analysis for Latvian. In: Frontiers in Artificial Intelligence and Applications. Volume 307: Human Language Technologies – The Baltic Perspective, pp. 112–119 (2018)

18. Ritter, A., Clark, S., Mausam, Etzioni, O.: Named entity recognition in tweets: an experimental study. In: Proceedings of the 2011 Conference on Empirical Methods in Natural Language Processing, pp. 1524–1534 (2011). https://www.aclweb.org/anthology/D11-1141

19. Souza, F., Nogueira, R., Lotufo, R.: Portuguese named entity recognition using BERT-CRF (2020). https://arxiv.org/abs/1909.10649

20. Vāravs, A., Salimbajevs, A.: Restoring punctuation and capitalization using transformer models. In: Dutoit, T., Martín-Vide, C., Pironkov, G. (eds.) SLSP 2018. LNCS (LNAI), vol. 11171, pp. 91–102. Springer, Cham (2018). https://doi.org/10.1007/978-3-030-00810-9_9

21. Vīksna, R., Skadiņa, I.: Large language models for Latvian named entity recognition. In: Human Language Technologies – The Baltic Perspective, pp. 62–69 (2020). https://doi.org/10.3233/FAIA200603

22. Virtanen, A., et al.: Multilingual is not enough: BERT for Finnish (2019). https://arxiv.org/abs/1912.07076

23. Znotiņš, A., Cīrule, E.: NLP-PIPE: Latvian NLP tool pipeline. In: Human Language Technologies – The Baltic Perspective, pp. 183–189 (2018). https://doi.org/10.3233/978-1-61499-912-6-183

24. Znotiņš, A., Barzdiņš, G.: LVBERT: transformer-based model for Latvian language understanding. In: Human Language Technologies – The Baltic Perspective, pp. 111–115 (2020). https://doi.org/10.3233/FAIA200610

Speech

Use of Speaker Metadata for Improving Automatic Pronunciation Assessment

Jose Antonio Lopez Saenz[✉][iD] and Thomas Hain[iD]

Speech and Hearing Research, Department of Computer Science,
University of Sheffield, Sheffield S10 2TN, UK
{jalopezsaenz1,t.hain}@sheffield.ac.uk

Abstract. Pronunciation assessment remains a subjective task which depends on a pronunciation reference hold as canonical. Whether a second language (L2) speaker is able to replicate said reference is decided by an assessor who perceives the identity of the sounds produced. It is known that the assessor has a bias caused by the perception of the speaker, hence the definition of a standard for L2 pronunciation is crucial in a formal assessment. In Computer Assisted Pronunciation Assessment (CAPA), the definition of a pronunciation standard for L2 is not trivial due to limited L2 data annotated for mispronunciations. Inspired on the assessor's bias, this work explores an alternative to a conventional Automatic Speech Recognition approach for CAPA by using speaker metadata along with acoustic observations for mispronunciation detection. A combination of Bidirectional Long-Short Memory with self-attention was used to detect pronunciation errors in short speech segments. It was found that the use of categorical metadata can have a positive effect in the classification of mispronounced segments depending on the sparsity and balance of the classes. It was also found that different assessors can be influenced differently by information about the speaker's linguistic background. The effect of the metadata was tested on data from Dutch children learners of English as L2 in schools across the Netherlands. The limited speaker diversity of the corpus made the task a challenge worth keep exploring.

Keywords: Pronunciation assessment · L2 learning · Speaker representation

1 Introduction

A speaker is considered to be proficient if they are capable of producing a sequence of vocal sounds assumed to be canonical for conveying a given word. The sounds hold as *correct* are not necessarily consistent across individuals as

Jose Antonio Lopez Saenz is a doctoral student from Programa de Becas en el Extranjero from CONACYT with the fellowship number 661687 at the University of Sheffield. We also want to thank ITSLanguage BV for the data facilitated.

© Springer Nature Switzerland AG 2021
L. Espinosa-Anke et al. (Eds.): SLSP 2021, LNAI 13062, pp. 61–72, 2021.
https://doi.org/10.1007/978-3-030-89579-2_6

speech is not a constant phenomena, yet some pronunciations are legitimized over others by being designated as a standard [13]. The assessment of a speaker is carried out by a listener whose perception decides the identity of the uttered sounds and consequently the proficiency of the speaker. The concept of the phoneme plays an important role in this process as it is often misunderstood as an acoustic building block for speech instead of the abstract phenomena it represents [15]. A listener would consider a segment is mispronounced if the perceived sounds differ from a pronunciation reference which is subject to personal experience and any previous training in the case of a formal assessment. Any variation from this reference also holds a bias from the listener which could be affected by factors characteristic of the perceived identity of the speaker [13]. Said effect becomes relevant in the pronunciation assessment of *second language* (L2) speakers since factors such as the presence of an accent or a pronunciation *close to natural* could be used as descriptors in scales of L2 pronunciation proficiency [10,20]. Skilled L2 speakers do not need to perfectly replicate a *native language* (L1) accent, although some minimal similarity to a pronunciation standard is required for communicating successfully. Trained assessors can distinguish speakers at both extreme points of a grading scale without much effort, it is the intermediate levels the ones subject to the most interpretations of the scale descriptors [10].

The effect the listener's bias has on assessment can be reduced, or made consistent across assessors, through proper training on the selected L2 pronunciation standard [10,22,24]. The definition of a pronunciation reference for *Computer Assisted Pronunciation Assessment* (CAPA) is not trivial, particularly for L2 speech as data resources are often limited. To this day, there is no publicly available L2 speech corpus marked for mispronunciation that serves as a reference for baselines or that could compare in size to other speech corpora used as a pronunciation reference such as the WSJCAM0 [16,17,23].

The use of conventional *Automatic Speech Recognition* (ASR) for CAPA is not ideal mainly due the need for a proper acoustic model (AM) for L2 speech [8]. The AM links phoneme identities with real sounds chosen to represent a pronunciation standard. Due to the scarcity of L2 data, acoustic models for CAPA are built using mainly L1 data. To improve the representation of L2 speech, the AM is adapted to L2 pronunciation to some extent via techniques such as Maximum Likelihood Linear Regression [7,8,12,23]. This work explores the use of speaker metadata for improving L2 *automatic mispronunciation detection* (AMD) as an alternative to an ASR approach with acoustic adaptation and to tackle the limited availability of L2 data resources. Based on the listener bias caused by the perception of the speaker [10,13,20], speaker information associated with their linguistic background is paired with acoustic observations as inputs. A combination of sequential encoding using a *Bidirectional Long Short-Term Memory* (BDLSTM) layer [18] and attention weights for saliency region selection is also introduced to perform AMD on short audio segments. This new implementation does not rely on precise timing information and learns a pronunciation reference using only L2 speech examples.

2 The ASR Approach for Automatic Pronunciation Assessment

A key task for CAPA is to detect the occurrence of a mispronunciation often using phoneme based scores to associate them with human annotations [8,12,23]. The scoring of phoneme segments is strongly dependant on the boundaries determined by a process of alignment, as it defines which observation samples contribute to the score. In [8], this problem is presented in a clear and clever manner under the scope of the Goodness of Pronunciation algorithm (GOP) [23], which remains used in CAPA to this day for scoring phoneme segments for mispronunciation [3,19]. The GOP scores correspond to the absolute log posterior probability of the expected phoneme p being present in the acoustic segment of interest $O^{(p)}$. The final score is obtained after normalizing the posterior over the length of the segment in number of frames $NF(p)$.

$$\mathbf{GOP}(p) = \frac{1}{NF(p)} \left| \log \left(\frac{p(O^{(p)}|p)\,P(p)}{\sum_{q \in Q} p(O^{(p)}|q)P(q)} \right) \right| \qquad (1)$$

In Eq. 1, $\mathbf{GOP}(p)$ is a competition between phonemes. If the phoneme produced shows the maximum likelihood for p in both numerator and denominator, the score lies under a decision threshold to be considered *correct*. If the score crosses said threshold, it is considered a mispronunciation. The thresholds are phoneme dependent and are chosen to better match human annotation.

Aside from the fundamental question of when does a phoneme starts and ends, other implications of GOP limit the score to reflect the true identity of the segment. The assumption of the phoneme p being present cannot guarantee an accurate alignment, which may result in a close to meaningless segmentation. The GOP also assumes that only one phoneme occurs during a segment, yet [8] showed this is not necessarily true, specially for non standard pronunciations such as L2 speakers and children. Additionally, the output of an ASR system trained mainly on L1 data will have difficulties modeling a L2 speaker who would likely use a set of phonemes different from the standard pronunciation of a target language.

3 A Segment Based Approach for Mispronunciation Detection

This work avoids the use of an ASR framework for pronunciation assessment. Instead, the strategy chosen consists in detecting the presence of a mispronunciation without requiring information about its precise location in time. The probability of a phoneme being mispronounced is estimated over a short acoustic segment $\mathbf{O}$ associated to the phoneme sequence $\mathbf{r} = \{r_i; i = 1, \dots, R\}$ considered to be a canonical pronunciation. The sequence $\mathbf{r}$ is also associated to a binary indicator of correctness $\mathbf{l} = \{l_i; i = 1, \dots, R\}$. Each $l_i = 1$ if the corresponding phoneme r_i is marked as correctly pronounced and $l_i = 0$ otherwise. A language learner could have produced a different phoneme sequence s not necessarily of the same length of r. It is still possible to allocate a correctness indicator q_j to any element in s, although the association of q with l could be complicated due the alignment of r and s. The probability of a pronunciation error in a segment O can be expressed as follows:

$$P(\text{error}|\mathbf{O}) = 1 - P(\mathbf{l} = 1|\mathbf{r}, \mathbf{O}) \qquad (2)$$

The canonical sequence r is known for each segment as it is assumed the assessment is performed on prompted speech. It is also assumed independence between errors for the sake of simplicity. Therefore, the probability of a segment free of errors becomes:

$$P(\mathbf{l} = 1|\mathbf{r}, \mathbf{O}) = \prod_i (l_i = 1|\mathbf{r}, \mathbf{O}) \qquad (3)$$

The equation above permits two types of equivalences, one that focus on a particular phoneme being present (Eq. 5) and one for information on a whole segment (Eq. 4).

$$P(l_i = 1|\mathbf{r}, \mathbf{O}) \equiv P(l_i = 1|r_i, \mathbf{O}_i) \qquad (4)$$
$$P(l_i = 1|\mathbf{r}, \mathbf{O}) \equiv P(l_i = 1|r_i, \mathbf{O}) \qquad (5)$$

The estimation of Eq. 4, in which $\mathbf{O}_i$ denotes the audio segment associated with phoneme r_i, is equivalent to the GOP as defined in Eq. 1. No timing information is required nor explicit manufacture of previous knowledge as long as the model has observed all variation. The model depends only on O and r to estimate the correctness indicators l. This direct estimation can be carried out without the need of a conventional ASR pipeline nor L1 data by implementing an attention-based sequence model (ABM) as defined in Sect. 4.

4   Attention Based Model

In the recent years, CAPA has relied less on conventional ASR pipelines by turning to *Long Short-Term Memory* (LSTM) to build End-to-End systems [4–6,9,26]. The architecture used in this work was inspired in how a human detects a mispronunciation by judging a phoneme sequence perceived and the benefits of modelling sequence dependencies using attention mechanisms found in [14].

A BDLSTM is chosen for spatial-sequential encoding. The BDLSTM has been proven useful for acoustic modeling and exploiting sequential relationships without using precise timing information [18,25]. In the notation of [2], the forward output hidden state $h_t^{\rightarrow}$ and the corresponding backward one $h_t^{\leftarrow}$ from the BDLSTM are concatenated ($\oplus$) as $h_t = [h_t^{\rightarrow}; h_t^{\leftarrow}]$ to form $\mathbf{h_O} = \{h_{O_{t_0}}, \ldots, h_{O_T}\}$ from the observation O. The encoding $\mathbf{h}_O$ is passed through a self-attention component. The objective is to make the model focus on particular spatio-temporal relationship which may be more relevant than others for a given assessor to associate a phoneme identity with the observed acoustic features. The model uses additive attention [2]. The attention component computes the energy $e_{i,t}$ over time as shown in Eq. 6, where v_i, W_i and V_i are weight matrices. The energy is normalized to generate the attention weights $\alpha_{i,t}$ establishing the relevance of each h_{o_t}. A context vector ctx is generated by multiplying element-wise ($\odot$) the attention weights with $\mathbf{h}_O$. The resulting vector is added to $\mathbf{h}_O$ creating a residual connection which eases the backpropagation of the gradient [11, 21]. The final encoding is passed through a normalisation layer [1] and a dropout layer with $p = 0.10$ for regularization.

$$e_{i,t} = v_i \odot \tanh(W_i^\top h_{o_t} + V_i^\top h_{o_t}) \tag{6}$$

$$\alpha_{i,t} = \frac{\exp(e_{i,t})}{\sum_{k=0}^{T}(e_{i,k})} \tag{7}$$

$$ctx = \alpha \odot \mathbf{h}_O \tag{8}$$

A deep feed-forward network (DFF) performs classification on the final sequential encoding. The output layer has an output for both correct and incorrect scenarios of every phoneme class marked in the data. The output configuration aims to exploit the most information available from the annotation and better deal with the possibility of the phoneme being considered absent. The output layer also allows that non canonical pronunciations contribute to the probability of an error being present.

The model does not require any linguistic information other than the vector of expected phonemes $\mathbf{r}$, which is not even sensitive to sequential order. The vector $\mathbf{r}$ is only involved during training and the scoring of a segment. The outputs corresponding to phonemes not present in $\mathbf{r}$ are set to zero to avoid underflow in the estimation of Eq. 3. The ABM is trained only using L2 data, forcing the model to infer a pronunciation standard particular to the observed speakers. Information about the speaker is also included as an observed variable to estimate Eq. 2 and help overcome the limited L2 speech data available. Non acoustic speaker factors related to their linguistic background are encoded into a representation λ, which is concatenated to $\mathbf{O}$ as additional constant dimensions over time. The use of metadata aims to better learn the assessors' perception towards the speakers. The construction of λ is specified in Sect. 6. The resulting ABM is illustrated in Fig. 1.

5 The Data

The corpus selected for this work is the ITSLanguage (ITSL) corpus from ITSLanguage BV, particularly the INA set. The data consists of prompted speech recordings from children and early teenagers learners of English as L2 in various schools across the Netherlands. The students recorded range in age, L2 proficiency level, L1, dialects and other characteristic that could be informative enough to construct a profile for pronunciation tendencies. The data was

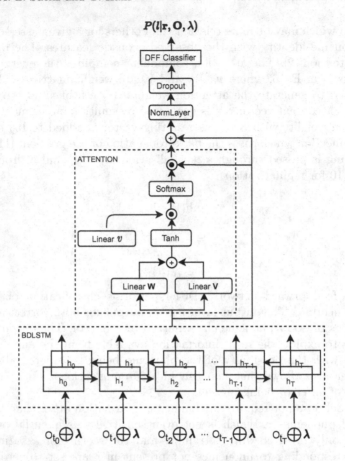

$$P(l|\mathbf{r}, \mathbf{O}, \lambda)$$

Fig. 1. Diagram for the ABM. The speaker representation λ is concatenated to the acoustic input $\mathbf{O}$, which ranges from frames t_0 to T. The phonemes $\mathbf{r}$ are used as a scoring condition rather than an actual input for the model

recorded in classrooms, hence a high level of environmental noise is present. Each student recorded themselves individually reading from a list of 193 short sentences and isolated words. A total of 80 h of data were recorded and a set of 6 h were chosen to form the INA set, which includes over 230 students. The INA set was forced aligned using a British English acoustic model and a multiple pronunciation dictionary [16]. Each example in INA was marked for mispronunciations at phoneme level by 3 trained phoneticians (a1, a2, a3). The amount of annotated data might not be considered large, yet the format of individual assessor marking allows testing for the presence of the listener bias. The agreement and cross correlation coefficients between the assessors is shown in Table 1.

A corpus like ITSL is distinctive due to the overlapping annotations at phoneme level. To the best of our knowledge there is no publicly available cor-

Table 1. Annotation agreement (A) and cross-correlation(CC) for each assessor pair of the INA set

vs.		A	CC
a1	a2	0.858	0.434
a2	a3	0.782	0.412
a3	a1	0.818	0.523

pus of L2 learners that also provides the individual annotation from multiple assessors for such amount of speakers.

6   Encoding of Speaker Factors

The students in the ITSL corpus filled up a questionnaire about their linguistic background, providing metadata that could be associated with their pronunciation tendencies. The factors selected were:

- Categorical:
    - **BP**: Dutch province or country of birth.
    - **DIAL**: Dutch dialect used.
    - **L1**: Native Language.
    - **SAL**: Self assessed English proficiency level.
    - **SCH**: School ID.
    - **YENG**: Years of formal studies of English as L2.
- Binary:
    - **MLH**: Multilingual household.
    - **NND**: Non-native Dutch speaker.
    - **NNP**: Non-native Dutch speaking parents.

The information was provided only by the students and was not corroborated by the teachers or tutors. For every factor from the metadata, each of the existent classes were anonymised and changed to numerical labels to protect the identity of the speakers. The number of classes in each factor listed above was kept down to the minimal possible to try make most of the information binary instead of multiple unbalanced classes.

7   Experiments

The encoded speaker metadata listed in Sect. 6 was used for training the ABM and test the effect on the model learning the human annotation. Individual speaker factors as well as combinations were used as λ. The ABM trained used a single BDLSTM layer of size 64, linear layers of size 128 for the attention module, and a DFF classifier with 4 hidden layers of size 1024. The models were evaluated on detecting segments that contain at least one mispronounced phoneme according to the human annotation; therefore the F1 score and Cohen's Kappa (K) were used as metrics. The performance of the models was tested on a consolidated annotation format obtained via maximum vote (MAX) as well as for each individual assessor in the corpus. The INA set was split designating 85% of the data for training while the rest was left for testing. The Train and Test sets contain no speaker overlap and hold a balance across sex, age and proficiency of the speakers.

An ABM trained with no λ was used as the baseline. Although the ABM does not require the time location of the phoneme segments, a forced alignment was run to designate the phonemes present in a segment. The alignment used DFF monophone state posteriors trained on WSJCAM0 and 46 h of ITSL data that do not overlap with the INA set. The same method for alignment was used in [16] and was proven useful for building phoneme discriminators for AMD on ITSL data. The phoneme posteriors from the alignment were also used to obtain GOP scores (Eq. 1) to perform AMD and compare the performance to the one of the baseline ABM. Only the GOP scores of the phonemes listed in the corresponding **r** for each **O** were considered for a fair contrast.

A sliding window of 0.5 s with a stride of 0.05 s was run over the INA recordings to define the segments **O**. Only phonemes which alignments were contained within at least 2 frames inside **O** were assumed as **r** to help the model overcome alignment errors. The segments showed a mean of 3.46 phonemes with a standard deviation of 1.54.

Perceptual Linear Predictor Coefficients (PLP) were chosen as acoustic features for their known noise robustness. The first 13 coefficients with their 1st and 2nd order differentials were extracted using a window size of 25 ms and a stride of 10 ms. The ABM was trained using the Adam optimizer and binary cross-entropy as loss criteria. The decision for declaring a segment as mispronounced was taken using a decision threshold representing the point of *equal error rate* (EER) on the train data.

8   Results and Discussion

The speaker factor combinations chosen aimed to help the model associate the metadata with pronunciation tendencies. The factors listed in Sect. 6 were not the only metadata tested. The performance metrics reported in this section correspond to the speaker factors which had the most effect in the performance of the model given the annotation reference.

The performance metrics of the models trained on MAX as well as the ones of the GOP are shown in Table 2 with the ABM baseline labeled as None. The ABM outperformed the alignment based method. The single speaker factors used for training the ABM do not show a positive effect greater than 0.01 on F1 and it is even less on K. Although the effect is not overwhelming, some factor pairs showed a greater improvement on the train set, particularly BP.L1 and BP.MLH. Overall, the models do not show the same effect on the Test set, which might be related to sparse and unbalanced metadata. For example, the factor L1 holds

6 classes, from which the only class overlapping between Train and Test data represents 96% of the data. The factor BP has a more even distribution with 14 classes from which the largest class proportion is 35% and multiple classes, yet not all of them, overlap across Train and Test sets.

Table 2. Performance results on ABM for the GOP and the ABM trained on the MAX consolidated annotation. The number of classes (C), F1 score and Cohen's Kappa (K) is shown for every factor combination used as λ

λ	C	Train		Test	
		F1	K	F1	K
GOP	–	0.5398	0.1624	0.4916	0.1510
None	–	0.7385	0.5351	0.6535	0.4213
BP	14	0.7422	0.5218	0.6564	0.4138
DIAL	10	0.7407	0.5391	0.6515	0.4199
L1	6	0.7293	0.5183	0.6483	0.4120
MLH	2	0.7333	0.5256	0.6510	0.4177
NND	2	0.7319	0.5230	0.6526	0.4186
NNP	2	0.7316	0.5225	0.6507	0.4161
SAL	5	0.7332	0.5255	0.6537	0.4237
SCH	7	0.7325	0.5242	0.6530	0.4232
YENG	21	0.7321	0.5234	0.6494	0.4183
BP.DIAL	31	0.7440	0.5452	0.6561	0.4257
BP.L1	24	0.7500	0.5563	0.6563	0.4269
BP.MLH	21	0.7476	0.5519	0.6603	0.4334
BP.NND	21	0.7346	0.5281	0.6527	0.4204
NND.MLH	4	0.7346	0.5281	0.6490	0.4136
NND.YENG	28	0.7358	0.5302	0.6515	0.4230
SCH.DIAL	25	0.7449	0.5469	0.6491	0.4185
BP.MLH.NNP	29	0.7463	0.5495	0.6570	0.4287
BP.NND.NNP	27	0.7451	0.5473	0.6562	0.4263
MLH.NND.YENG	34	0.7370	0.5324	0.6543	0.4262
SCH.DIAL.YENG	104	0.7465	0.5498	0.6506	0.4236

Another example worth mentioning is SCH as it is probably the most even distributed factor due to schools having a similar number of students. The use of SCH alone did not cause any improvement in the metric, yet it managed to increase the positive effect of DIAL by pairing up with it. The positive effect of SCH.DIAL increased even more by concatenating YENG; however, the F1 decreased on the Test set for both SCH.DIAL pairings as these represent even less overlapping classes across sets.

The models trained on the individual assessors put on evidence the annotation variability reduced in MAX. Each assessor holds a different pronunciation standard as seen in the results from Table 3. The performance of the models was consistent across λ factors for every assessor, showing how dependant CAPA is on the reference hold by the markers. Most of the improvement in AMD occurred for a3, which coincidentally has the baseline with the highest metrics. The behaviour of the a3 models might reflect a relatively major consistency in the marking which benefited from λ. Similar to MAX, most of the combinations involving BP had a positive effect on the a3 models.

Speaker factor classes with a relatively small occurrence and no overlap between data sets will likely be considered noisy data by the model. The observed distribution of the metadata is consistent with a speaker population with less than 4% of the data produced by speakers of different L1.

Table 3. Performance results on ABM for the GOP and the ABM trained on assessors a1, a2 and a3. The F1 score and Cohen's Kappa (K) is shown for every factor combination used as λ

λ	a1				a2				a3			
	Train		Test		Train		Test		Train		Test	
	F1	K	F1	K	F1	K	F1	K	F1	K	F1	K
GOP	0.5574	0.1757	0.5151	0.1703	0.4158	0.1292	0.3642	0.1216	0.6659	0.2116	0.6437	0.2246
None	0.7283	0.5021	0.6450	0.3857	0.6167	0.4492	0.5091	0.3338	0.8342	0.5806	0.7875	0.4879
BP	0.7236	0.4932	0.6416	0.3817	0.6127	0.4430	0.5034	0.3267	0.8326	0.5768	0.7896	0.4908
DIAL	0.7233	0.4926	0.6428	0.3843	0.6075	0.4349	0.5012	0.3223	0.8315	0.5742	0.7885	0.4901
L1	0.7230	0.4920	0.6419	0.3820	0.6244	0.4610	0.5068	0.3314	0.8307	0.5724	0.7886	0.4902
MLH	0.7223	0.4908	0.6422	0.3830	0.6118	0.4417	0.5022	0.3232	0.8412	0.5972	0.7887	0.4915
NND	0.7227	0.4915	0.6434	0.3837	0.6269	0.4648	0.5108	0.3371	0.8317	0.5746	0.7891	0.4908
NNP	0.7231	0.4922	0.6400	0.3804	0.6277	0.4661	0.5066	0.3316	0.8327	0.5769	0.7882	0.4883
SAL	0.7256	0.4969	0.6416	0.3853	0.6168	0.4494	0.5046	0.3295	0.8320	0.5754	0.7892	0.4921
SCH	0.7211	0.4885	0.6431	0.3878	0.6110	0.4404	0.5006	0.3233	0.8322	0.5759	0.7875	0.4904
YENG	0.7240	0.4941	0.6406	0.3847	0.6135	0.4442	0.5053	0.3317	0.8318	0.5749	0.7879	0.4912
BP.DIAL	0.7295	0.5044	0.6470	0.3925	0.6161	0.4482	0.5057	0.3305	0.8401	0.5945	0.7884	0.4913
BP.L1	0.7260	0.4977	0.6480	0.3929	0.6212	0.4561	0.5039	0.3277	0.8416	0.5981	0.7915	0.4947
BP.MLH	0.7248	0.4955	0.6458	0.3892	0.6179	0.4510	0.5058	0.3300	0.8454	0.6069	0.7881	0.4895
BP.NND	0.7244	0.4947	0.6460	0.3885	0.6172	0.4499	0.5034	0.3267	0.8379	0.5894	0.7897	0.4923
NND.MLH	0.7242	0.4944	0.6455	0.3890	0.6180	0.4512	0.5053	0.3291	0.8409	0.5963	0.7929	0.5000
NND.YENG	0.7242	0.4944	0.6462	0.3950	0.6164	0.4488	0.5031	0.3290	0.8345	0.5813	0.7882	0.4945
SCH.DIAL	0.7368	0.5181	0.6472	0.3956	0.7449	0.5469	0.6491	0.4185	0.8435	0.6026	0.7900	0.4971
BP.MLH.NNP	0.7244	0.4947	0.6413	0.3788	0.7449	0.5469	0.6491	0.4185	0.8414	0.5977	0.7894	0.4895
BP.NND.NNP	0.7256	0.4969	0.6466	0.3906	0.6167	0.4492	0.4988	0.3191	0.8404	0.5953	0.7870	0.4855
MLH.NND.YENG	0.7265	0.4986	0.6451	0.3926	0.6194	0.4534	0.5070	0.3333	0.8331	0.5781	0.7882	0.4905
SCH.DIAL.YENG	0.7249	0.4958	0.6434	0.3921	0.6167	0.4492	0.5043	0.3320	0.8304	0.5717	0.7855	0.4885

Aside from the speakers selected not being notably diverse and the fact that the metadata was collected without any proof of the students' claims, certain combination of speaker factors still had a positive effect on the performance metrics of a model trained without any L1 data nor conventional acoustic adaptation. More combinations for λ have yet to be explored considering that speaker factors with a more even distribution could have a greater positive effect than factors with notably sparse and unbalanced classes.

9   Conclusions

This work presented a method to improve AMD by using speaker metadata. A segmental based approach was used to train an ABM instead of a conventional ASR pipeline which may depend on assumed L1 and L2 references. The ABM outperformed a standard GOP implementation and learned a pronunciation reference using only L2 speech data and the annotation available. A speaker representation λ was built using different combinations of speaker metadata which was concatenated to the acoustic inputs. Different speaker metadata showed a positive effect in the performance metrics of the AMD given the sparsity and class balance of λ. The findings in this work were obtained using data from which most of the speakers share L1, dialects and various linguistic characteristics. A more balanced and diverse L2 speech corpus can encourage CAPA research to divert from conventional ASR approaches and assumptions about pronunciation references.

References

1. Ba, J.L., Kiros, J.R., Hinton, G.E.: Layer normalization (2016)
2. Bahdanau, D., Cho, K.H., Bengio, Y.: Neural machine translation by jointly learning to align and translate. In: Proceedings of the 3rd International Conference on Learning Representations, ICLR 2015 - Conference Track Proceedings, pp. 1–15 (2015)
3. Chen, L., Gao, Q., Liang, Q., Yuan, J., Liu, Y., China, L.I.S.: Automatic scoring minimal-pair pronunciation drills by using recognition likelihood scores and phonological features. In: SLaTE, pp. 25–29 (2019)
4. Chen, L., Tao, J., Ghaffarzadegan, S., Qian, Y.: End-to-end neural network based automated speech scoring. In: 2018 IEEE International Conference on Acoustics, Speech and Signal Processing (ICASSP), pp. 6234–6238. IEEE (2018)
5. Chen, L., et al.: End-to-end neural network based automated speech scoring Midea America Corporation, 250 W Tasman Dr, San Jose, CA 95134, USA Robert Bosch Corporation, 4005 Miranda Ave, Palo Alto, CA 94304, USA Educational Testing Service (ETS), 90 New Montgomer. In: 2018 IEEE International Conference on Acoustics, Speech and Signal Processing (ICASSP), pp. 6234–6238 (2018)
6. Cheng, S., Liu, Z., Li, L., Tang, Z., Wang, D., Zheng, T.F.: ASR-free pronunciation assessment. arXiv pp. 3047–3051 (2020)
7. Chu, W., Liu, Y., Zhou, J.: Recognize mispronunciations to improve non-native acoustic modeling through a phone decoder built from one edit distance finite state automaton. In: INTERSPEECH, pp. 3062–3066 (2020)
8. Dudy, S., Bedrick, S., Asgari, M., Kain, A.: Automatic analysis of pronunciations for children with speech sound disorders. Comput. Speech Lang. 50, 62–84 (2018)
9. Fu, K., Lin, J., Ke, D., Xie, Y., Zhang, J., Lin, B.: A full text-dependent end to end mispronunciation detection and diagnosis with easy data augmentation techniques (2021)

10. Harding, L.: What do raters need in a pronunciation scale? The user's view. In: Isaacs, T., Trofimovich, P. (eds.) Second Language Pronunciation Assessment: Interdisciplinary Perspectives, chap. 2, pp. 12–34. Multilingual Matters/Channel View Publications (2017)
11. He, K., Zhang, X., Ren, S., Sun, J.: Deep residual learning for image recognition. In: Proceedings of the IEEE Conference on Computer Vision and Pattern Recognition, pp. 770–778 (2016)
12. Huang, G., Ye, J., Shen, Y., Zhou, Y.: A evaluating model of English pronunciation for Chinese students. In: 2017 IEEE 9th International Conference on Communication Software and Networks (ICCSN), pp. 1062–1065. IEEE (2017)
13. Lindemann, S.: Variation or 'error'? perception of pronunciation variation and implications for assessment. Second language pronunciation assessment, p. 193 (2017)
14. Milner, R., Jalal, M.A., Ng, R.W., Hain, T.: A cross-corpus study on speech emotion recognition. In: 2019 IEEE Automatic Speech Recognition and Understanding Workshop (ASRU), pp. 304–311. IEEE (2019)
15. Moore, R.K., Skidmore, L.: On the use/misuse of the term'phoneme'. arXiv preprint arXiv:1907.11640 (2019)
16. Nicolao, M., Beeston, A.V., Hain, T.: Automatic assessment of English learner pronunciation using discriminative classifiers. In: 2015 IEEE International Conference on Acoustics, Speech and Signal Processing (ICASSP), pp. 5351–5355. IEEE (2015)
17. Robinson, T., Fransen, J., Pye, D., Foote, J., Renals, S.: WSJCAMO: a British English speech corpus for large vocabulary continuous speech recognition. In: 1995 International Conference on Acoustics, Speech, and Signal Processing, vol. 1, pp. 81–84. IEEE (1995)
18. Sak, H., Senior, A., Beaufays, F.: Long short-term memory based recurrent neural network architectures for large vocabulary speech recognition (2014)
19. Sudhakara, S., Ramanathi, M.K., Yarra, C., Ghosh, P.K.: An improved goodness of pronunciation (GoP) measure for pronunciation evaluation with DNN-HMM system considering hmm transition probabilities. In: INTERSPEECH, pp. 954–958 (2019)
20. Trofimovich, P., Isaacs, T.: Second language pronunciation assessment: a look at the present and the future. Second Language Pronunciation Assessment, p. 259 (2017)
21. Vaswani, A., et al.: Attention is all you need. In: Advances in Neural Information Processing Systems, pp. 5998–6008 (2017)
22. Wei, J., Llosa, L.: Investigating differences between American and Indian raters in assessing TOEFL iBT speaking tasks. Lang. Assess. Q. 12(3), 283–304 (2015)
23. Witt, S.M., Young, S.J.: Phone-level pronunciation scoring and assessment for interactive language learning. Speech Commun. 30(2–3), 95–108 (2000)
24. Witteman, M.J., Weber, A., McQueen, J.M.: Tolerance for inconsistency in foreign-accented speech. Psychon. Bull. Rev. 21(2), 512–519 (2014). https://doi.org/10.3758/s13423-013-0519-8, http://link.springer.com/10.3758/s13423-013-0519-8
25. Zeyer, A., Doetsch, P., Voigtlaender, P., Schluter, R., Ney, H.: A comprehensive study of deep bidirectional LSTM RNNS for acoustic modeling in speech recognition. ICASSP, IEEE International Conference on Acoustics, Speech and Signal Processing - Proceedings pp. 2462–2466 (2017). doi: https://doi.org/10.1109/ICASSP.2017.7952599
26. Zhang, L., et al.: End-to-end automatic pronunciation error detection based on improved hybrid ctc/attention architecture. Sensors 20(7), 1809 (2020)

Augmenting ASR for User-Generated Videos with Semi-supervised Training and Acoustic Model Adaptation for Spoken Content Retrieval

Yasufumi Moriya[✉] and Gareth J. F. Jones

ADAPT Centre, School of Computing, Dublin City University, Dublin 9, Ireland
{yasufumi.moriya,gareth.jones}@adaptcentre.ie

Abstract. We present an investigation into the use of semi-supervised training and content genre adaptation for improved automatic speech recognition (ASR) of diverse user-generated videos in the task of spoken content retrieval (SCR). Previous work has successfully applied semi-supervised training in single domain ASR tasks. Our focus is on the exploration of the effective use of semi-supervised training of ASR systems for transcription of the spoken content stream of user-generated video data in varied domains and acoustic noise conditions for use in SCR systems. We examine all elements of ASR system development including: data segmentation, data selection, genre labels, acoustic modelling and language modelling using semi-supervised training. We evaluate its effectiveness for ASR and a known-item SCR task using the Blip100000 multimedia collection. Our baseline hybrid ASR system trained out-of-domain produced WERs 31.27% and 44.69% on dev and test sets, respectively. By introducing the techniques outlined above, the WERs are reduced to 26.82% and 39.21% respectively. The improved transcripts increased mean reciprocal rank (MRR) results for the SCR task from 15.59% to 39.38% on dev and 20.98% to 37.23% on test sets.

Keywords: Spoken content retrieval · Speech recognition · User generated data · Semi supervised training · Content genre adaptation

1 Introduction

The growing amount of digital multimedia content such as user-generated videos and podcasts, now widely available on the Internet, is increasing the importance of effective Spoken Content Retrieval (SCR) systems. SCR systems generally operate using speech transcripts created using automatic speech recognition (ASR). It is known that SCR effectiveness is generally impacted by high word error rates (WERs), e.g. greater than 30% [4]. High error rates in speech transcripts can cause a "mismatch" between user search queries and document transcripts, even if the documents are highly relevant to the queries. State-of-the-art ASR systems show very low WERs for well controlled transcription tasks

© Springer Nature Switzerland AG 2021
L. Espinosa-Anke et al. (Eds.): SLSP 2021, LNAI 13062, pp. 73–84, 2021.
https://doi.org/10.1007/978-3-030-89579-2_7

such as for the Wall Street Journal (WSJ) and LibriSpeech corpora [2,5]. However, transcription of uncontrolled, highly varied user-generated speech remains a challenging ASR task often with high WERs.

The key challenges of transcribing user-generated spoken video arises from the highly varied speaker characteristics (adult, child and non-native speaker), speaking styles (scripted, formal and informal interviews, sports and video game commentary, and casual conversations), and acoustic conditions (background music, loud audience, applause, and street noise). The goal of our work reported here is to develop a multi-domain ASR system suitable for such challenging user-generated videos, and evaluate its effectiveness both in terms of WER and for an SCR task.

In this paper, we investigate the use of semi-supervised training and video genre tag for development of multi-domain ASR. While past work on multi-domain ASR [8] assumed the existence of manual transcripts or user-uploaded captions, our work focuses on the use of untranscribed speech for semi-supervised training. Further, our ASR system trained using a semi-supervised approach is evaluated in the context of SCR, which has not been the end goal of existing research on semi-supervised ASR training [7,19]. In semi-supervised settings, a seed ASR system is used to generate transcripts or decoded lattices of the data from which a new ASR system is created. The absence of a requirement for manually transcribed training data makes semi-supervised training an attractive option for addressing the challenges of transcribing diverse user-generated videos. User-generated videos are often accompanied by a video genre tag uploaded by the user. We exploit the genre tag information for acoustic model adaptation [1,15]. Our hypothesis is that user-generated videos with the same genre tag will tend to contain the same type of acoustic events (e.g., applauding in "conference" and loud audience in "sports").

The rest of the paper is organised as follows. Section 2 examines ASR system development using untranscribed data. Section 3 presents acoustic model adaptation using content genre, followed by experimental investigations in Sect. 4 and Sect. 5. Section 6 provides concluding remarks of this paper.

2   Semi-supervised Acoustic and Language Modelling

The overall goal of our investigation is to develop an ASR system which improves SCR effectiveness. We hypothesize that reducing WERs in ASR by optimising data segmentation, data selection, acoustic modelling and language modelling using untranscribed data can help to achieve this. Figure 1 shows a flowchart for the application of the semi-supervised approach to acoustic model and language model training in ASR.

Data Segmentation. Natural audio generally consists of a mixture of speech utterances and other audio activities. The curated speech corpora used in existing ASR research are typically pre-segmented into speech utterances. We wish to make use of untranscribed raw video data in our work in which boundaries between speech and other audio data are unknown. We thus seek to identify

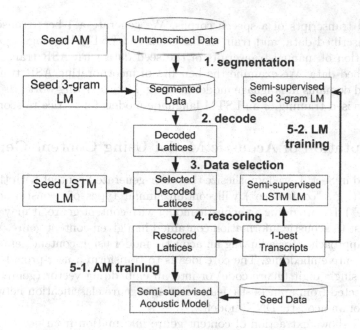

Fig. 1. Flowchart of semi-supervised training for acoustic and language models.

regions of speech using VAD and use this to segment the video data. To explore the usefulness of this VAD step, we compare VAD segmented data with simple equal sized data segments. To do this, we compare ASR WER for a system trained on VAD segments with one trained on equal sized segments.

Data Selection. Data selection is used to select segments of untranscribed data which are likely to lead to sufficiently accurate ASR for semi-supervised training of an acoustic model and a language model. For our investigation, we use the segment level confidence score described in [22]. This is computed by taking the average of posterior probabilities of speech segments decoded by a seed ASR system. Either VAD or equally segmented untranscribed data below the set confidence level is then excluded from the training data.

Acoustic Model. For acoustic model training using untranscribed data, we use the recently proposed semi-supervised lattice-free maximum mutual information (LF-MMI) training method [7]. LF-MMI is discriminative training method where the training objective is to predict a sequence of phone labels as a whole, rather than individual phones in an utterance. In semi-supervised settings, several paths of a decoded lattice of untranscribed data are considered to be the target. When these paths contain lower probabilities (i.e., the seed system is not confident in its prediction), its impact on a new acoustic model is smaller.

Language Model. As shown in Fig. 1, a seed n-gram language model is used to decode untranscribed data and a seed LSTM language model is used to re-score decoded lattices of untranscribed data [21]. These language models are trained on manual transcripts of a speech corpus. We generate a 1-best transcript of the untranscribed data, and train new n-gram and LSTM language models on a combination of manual transcripts of the seed data with ASR transcripts of untranscribed data. We examine the benefits of incorporating ASR transcripts from varied domains in language model training. Little existing work has studied semi-supervised training of an LSTM language models for lattice re-scoring.

3   Adaptation of Acoustic Model Using Content Genre

As outlined in Sect. 1, we hypothesize that user-generated content with the same genre tag (e.g., "conference") will contain similar types of acoustic activities ("applause"). By providing an acoustic model with content genre, it may become more robust to acoustic information contained in a given content genre. We propose two approaches for adapting an acoustic model using content genre: genre code and genre embedding. The core idea is to transform a user-provided genre tag into a single digit (genre code) or into an embedding vector (genre embedding) extracted from an acoustic feature using a genre classification network for the input of an acoustic DNN model.

Figure 2 shows extraction of content genre information from user-generated video data. A genre code is generated by converting each of the unique genre tags to a digit. This is similar to the domain ID used in [13], however, we apply genre codes in the semi-supervised settings. This is treated as adaptation information and concatenated with an acoustic feature vector.

A more sophisticated approach is inspired by an x-vector system designed for speaker recognition [17]. The x-vector, which is useful to identify a speaker from speech, can be extracted from a DNN model trained to classify a speaker given an acoustic feature vector. The x-vector extractor consists of five layers operated on several speech frames with specified context, followed by a statistics pooling layer and two layers operated on speech segments. The x-vector is extracted from the first segment layer of the extractor. To extract genre embedding, we train a DNN model with the same structure as the x-vector extractor which can classify content genre given an acoustic feature vector, genre embedding can be extracted from the first segment layer of the extractor. This embedding is concatenated with an acoustic feature vector as input to an acoustic model.

4   ASR Experiments

4.1   Creation of Manual Transcripts for Blip10000

To evaluate our proposed use of semi-supervised training and content genre adaptation on diverse user-generated videos we use the Blip10000 corpus [16]. The Blip10000 corpus is a collection of user-generated videos of diverse qualities and genres crawled from the internet. It contains 14,838 videos (3,288 h) released under Creative Commons. The corpus is partitioned into 5,288 videos for dev set and 9,550 videos for test set. The Blip10000 corpus contains videos of 26 different

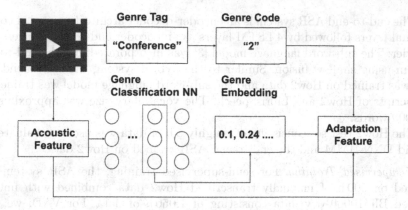

Fig. 2. Generation steps of genre code and genre embedding for content genre adaptation of an acoustic model.

genres; its content includes materials such as vlogs, conferences, street interviews, semi-professional broadcasts, technology reviews and so on. The spoken language is mainly English, but non-English videos can be found. To use blip10000 for ASR research, we created manual transcripts of a subset of data as follows: 670 videos of dev set (20 h) and 566 videos of test set (15 h)[1]. This amount of manually transcribed data enables us to study ASR behaviour on a much wider range of data than is typically the case in ASR research. Videos were selected from shorter ones to increase the number of documents and the diversity of content. The selected videos were manually transcribed by crowd-sourcing using Amazon Mechanical Turk (AMT).

4.2 Experimental Setup

Baseline Systems. We built two baseline systems: a hybrid DNN-HMM system using Kaldi [10] and an end-to-end ASR system using espresso [20].

The hybrid system consists of a DNN acoustic model, an n-gram language model and an LSTM language model for lattice re-scoring. The acoustic DNN model consists of 17 time-delayed layers with 1,024 units each trained using LF-MMI [11]. The n-gram is a 3-gram model built using the SriLM toolkit [18]. The LSTM language model consists of 2 layers of LSTM layers with 256 units each. For the hybrid system, we trained one acoustic model on approximately 500 h of How2 data containing instruction videos [14] and another model on 960 h of LibriSpeech audio-book data [9]. The How2 data is a collection of user-generated instructional videos, its acoustic conditions are more similar to those of Blip10000 than LibriSpeech. Nevertheless, the domain of How2 videos is limited to instruction and most of the How2 videos contain a single speaker, whereas the number of speakers varies in the Blip10000 data. The n-gram and LSTM language models were trained on manual transcripts of How2 and 960 h transcripts of the LibriSpeech corpus.

---

[1] We plan to make these manual transcripts publicly available.

. The end-to-end ASR system is an encoder-decoder architecture with 4 convolutional layers followed by 4 LSTM layers as an encoder and 4 LSTM layers as a decoder. The sub-word language model [3] was incorporated into the end-to-end system using shallow fusion. Similar to the hybrid system, the end-to-end system was trained on How2 data and the sub-word language model was trained on transcripts of How2 and LibrisSpeech. The vocabulary size was approximately 100,000 words.

The How2 test set consisting of roughly 5 h of data was used to evaluate the hybrid DNN-HMM and the end-to-end ASR trained on How2 data.

Semi-supervised Training. For semi-supervised training, the ASR system was trained on 500 h of manually transcribed How2 data combined with untranscribed Blip10000 dev data consisting of 1,050 h of data. For VAD, we used the NeMo toolkit [6] trained on the Google Speech Commands and Freesound datasets. This tool is claimed to classify speech and non-speech frames with 99% accuracy[2]. Untranscribed data were split into segments when non-speech frames were longer than 2 s. We added 0.5 of non-speech frames to the beginning and end of each segment to avoid abrupt cut-offs. Equal segments were created by segmenting untranscribed data into 30 s chunks with 5 s overlap with adjacent segments. Segments of 30 ss were quick to process with the seed system, but 5 ss of overlap ensures no abrupt cut-offs. We found that rejecting segments of untranscribed data with confidence score lower than 80% was the most effective both for VAD and equal segmentation.

For content genre adaptation of the acoustic model, we generated a genre code by transforming a genre tag of each Blip10000 video into a unique digit (e.g., 1: "technology", 2: "documentary"). Since How2 videos are not classified into different genres and all are instruction videos, How2 speech segments share the same genre code (i.e., 0). In the case of genre embedding, the genre tags were used for supervised training of an genre embedding extractor which classified a genre tag given acoustic features. Following the original paper on xvector [17], we trained a DNN genre embedding extractor with the embedding size set to 512. Genre embedding was the output of the first segment layer of the extractor. Along with the untranscribed dev set of Blip10000, audio from How2 data was added to the training data of the extractor. Acoustic features used in this paper are 40 dimensional MFCCs.

4.3   Experimental Results

Baseline Results. Table 1 shows WER results on the How2 test set and on the Blip10000 dev and test sets for the baseline hybrid ASR system and the end-to-end system. The results show that an acoustic model trained on How2 (instruction videos) is more suitable than LibriSpeech (audiobooks). There is a relatively small difference of 3.5% in WERs between the hybrid and the end-to-end system on the How2 test set. However, there is a much larger gap between the hybrid and end-to-end systems between the Blip10000 dev and test set. Two explanations are as follows. Firstly, the How2 data is spoken videos, but its domain is

---

[2] https://ngc.nvidia.com/catalog/models/nvidia:vad_matchboxnet_3x1x1.

Table 1. WERs of the baseline hybrid DNN-HMM systems and the end-to-end system on How2 test set and Blip10000. (lr) the seed LSTM LM re-scored lattices of the Blip10000 dev and test sets

	how2 test	blip dev	blip test
hybrid How2	13.19	31.27	44.69
hybrid How2(lr)	N/A	28.92	42.23
hybrid Libri	N/A	35.94	51.42
end2end	16.77	63.05	78.23

Table 2. WER results of semi-supervised training for acoustic and language modelling. The top three rows show results of semi-supervised acoustic model, seed LSTM re-scoring lattices of untranscribed data and seed LSTM re-scoring lattices of evaluation data. The bottom two rows show results of enhancing n-gram and LSTM language model with automatic transcripts of untranscribed data.

	dev		test	
	eq	vad	eq	vad
AM-semisup	29.85	30.06	42.65	43.02
+seedLSTM-rescore	29.54	29.59	42.39	42.50
+seedLSTM-eval	27.70	27.67	40.44	40.42
seedLSTM-rescore+ngram	28.99	28.89	41.61	41.67
+semisupLSTM-eval	**27.28**	**27.07**	**39.68**	**39.71**

limited to instructional videos. Both the ASR systems experienced worse WERs on the Blip10000 dev and test due to the systems being trained on the out-of-domain data. Secondly, the end-to-end system produced much worse results than the hybrid system, most likely because it requires a much greater amount of training data to achieve satisfactory recognition output. For example, recent work on an end-to-end system used 162,000 h of transcribed data or data with user-uploaded captions which is roughly equal to 18.5 years of data [8]. For the remainder of our experiments, our baseline system is the hybrid system trained on How2 audio, since this system produced the best WERs. All results after this section are only evaluated on the transcribed Blip10000 dataset.

Semi-supervised Training Results. Table 2 shows WER results achieved using semi-supervised training for acoustic and language models. Note that lattice re-scoring can be applied to lattices of untranscribed data and to lattices of evaluation data. Training an acoustic model on re-scored lattices of untranscribed data led to a 0.3-0.5% reduction in WERs. Re-training the n-gram on manual transcripts and 1-best transcripts of untranscribed data led to 0.3-0.8% reduction in WERs. Applying the new LSTM language model to re-scoring the lattices of the evaluation set produced further gain in WERs.

Table 3. WER results of semi-supervised trained systems using content genre adaptation. Top rows (1–3) show WERs without lattice rescoring by the seed LSTM language model, while middle rows (4–6) show WERs by the seed LSTM language model applied to the evaluation set. Bottom row shows the n-gram and the LSTM language model trained on the 1-best transcripts of unsupervised data and transcripts of seed data to the genre embedding system.

	dev		test	
	eq	vad	eq	vad
noadapt	29.54	29.59	42.39	42.50
genreCode	29.17	29.56	42.27	42.58
genreEmb	29.10	29.22	41.69	41.99
+seed-LSTM-eval				
noadapt	27.70	27.67	40.44	40.42
genreCode	27.42	27.63	40.25	40.45
genreEmb	27.29	27.35	39.91	39.95
+semisupLMs				
genreEmb	**26.82**	**26.94**	**39.21**	**39.29**

We found that there was no difference between the ASR systems trained on equal segments and VAD segments. This was surprising since applying VAD for data segmentation was expected to generate cleaner segments containing less background noise and silence. After rejecting segments lower than 80% confidence, segments created by equal segmentation retained 732 h of data, while segments created by VAD segmentation were 415 h. There were 2,874 videos in equal segments while 2,331 videos in VAD segments. 634 videos were observed only in equal segments, while 91 were only in VAD segments. This shows that output of VAD segments was almost a subset of output of equal segments. The VAD system used to create segments was trained on different data domains (Sect. 4.2) from Blip10000. This may explain why the VAD system filtered out too many speech frames from the untranscribed data.

Content Genre Adaptation Results. Table 3 shows results of using content genre for acoustic model adaptation. Results of "noadapt" correspond to +seedLSTM-rescore and +seedLSTM-eval in Table 2. A simple addition of genre code to an acoustic feature vector led to a small gain in WERs. Using the classifier to generate genre embedding resulted in the best WER among the systems 29.10% on dev and 41.69% on test without lattice rescoring, and 27.29% on dev and 39.91% on test sets with lattice rescoring. Further reduction in WERs of 0.45% on dev and 0.70% on test was obtained by decoding Blip dev and test using the n-gram re-trained on combination of the original training data with semi-supervised data and re-scoring lattices using the LSTM language model re-trained on combination of the original training data with semi-supervised data.

Overall, our experiments demonstrate that both semi-supervised training and content genre adaptation of an acoustic model can be effective for transcription of highly varied user-generated videos. The best configuration is to use equal segmentation of raw video data with removal of segments with confidence score less than 80%, addition of 1-best transcripts of untranscribed Blip10000 dev data to the n-gram and LSTM language model training, and genre embedding adaptation of the acoustic model together with semi-supervised training.

5 SCR Experiments

5.1 Creation of Known-Item Queries for Blip10000

To evaluate utility of the transcripts created using our alternative ASR systems for SCR we created 15 known-item search queries for dev set and 35 queries for test set using AMT. A known-item search seeks to re-find a previously observed relevant item. 15 documents from transcribed dev set and 35 documents from transcribed test set were randomly selected. AMT Workers were presented with the manual transcript of each document and a video. The workers were asked a question "Suppose you would like to find this video content on YouTube or another video sharing platform, enter minimum 3 words you would put in the search box". The workers were asked to create queries for not more than 3 documents to ensure that the query set was created by a diverse range of workers

5.2 Experimental Setup

We created search indexes of Blip10000 dev and test from ASR transcripts of the baseline ASR system (hybrid How2) in Table 1 and the augmented ASR system using semi-supervised acoustic model, language model and genre embedding (genreEmb) in Table 3. The dev document collection and the test document collection were indexed separately. In addition, the indexes of dev and test set were created using the manual transcripts described in Sect. 4.1. Since only 670 videos of dev set and 566 videos of test set have been manually transcribed, the indexed collections here were smaller than the actual document collections. The manually transcribed indexes were used as the oracle. The standard probabilistic BM25 information retrieval model was used to rank documents for each search query [12]. BM25 computes a relevance score given a document and a query by analysing frequency and inverse document frequency (IDF) of each query term in a document.

We report results for the known-item search task using the standard Mean Reciprocal Rank (MRR) metric conventionally used for known-item search tasks, where each query has a single relevant document. MRR is defined as follows:

$$MRR = \frac{1}{N} \sum_{i=1}^{N} \frac{1}{rank_i} \tag{1}$$

where N is the number of user queries and $rank_i$ is the rank of the document relevent for the ith query.

Table 4. MRR results for known-item search using BM25 applied to the baseline and to the genre embedding transcripts.

	baseline	genreEmb	% change	oracle
dev	15.59	39.38	+152.6%	82.95
test	20.98	37.23	+77.45%	76.68

5.3   Experimental Results and Analysis

Table 4 summarises MRR scores using the ASR baseline and augmented transcripts. The augmented ASR transcripts produced very large relative improvements in MRR of 152.6% on dev and by 77.45% on test sets over the baseline transcripts. Improvement of the MRR scores are statistically significant ($p < .05$). This demonstrates that the 4–4.5% improvement of WERs gained from semi-supervised training and acoustic model adaptation led to significant improvement in SCR effectiveness for this task. For the dev set, MRR scores for 8 queries improved for 3 queries decreased and for the remaining 4 queries did not change. For test set MRR scores improved for 24 queries, decreased for 6 queries, and did not change for remaining 5 queries. These results are though still far below results achieved using error free manual transcripts. Despite being error free, manual transcripts do not achieve perfect results since queries can still score higher against non-relevant documents in cases where using the BM25 algorithm when they match the query better than the relevant item.

Table 5 shows an analysis of success and failure cases using the augmented ASR transcripts for the known-item search. The three queries 2, 5 and 38 show dramatic improvement in MRR score. This is brought about by the improvement of the ASR transcripts. We note that the augmented ASR system correctly transcribed "Julia Morris" for query 2, while the baseline system replaced it for "Egeria Moa". Similarly, the surname "Marcy" was replaced for "Mercy" by the baseline system for query 5 and "Hansel's Affair" was replaced for "Humps Hills" for query 38. This demonstrates that semi-supervised training and acoustic model adaptation using genre embedding help to improve search effectiveness. The more interesting cases are the bottom three queries at Table 5. Despite improvement in WER of documents relevant to the queries, MRR scores were not better. For query 10 is due to the fact that transcription of the proper noun "Mail Chimp" did not succeed. The retrieval model de-ranked the relevant document for query 24, since the augmented ASR system correctly transcribed the term "revamp" in another document irrelevant to the query and this document was ranked higher than the relevant document. Retrieval failure for query 47 occurred because the baseline transcript of the relevant document contained the term "Rocco", while the augmented transcript did not. These failure cases show the importance of correctly transcribing named entities for the search task, which could not be addressed by our enhancements to ASR system.

Table 5. Example queries for which MRR score increased or decreased when using augmented ASR transcripts. MRR scores of queries are shown with WERs of the relevant document.

queryID	query	baseline		genreEmb	
		MRR	WER	MRR	WER
2	Ancient craft film, Julia morris gallery ...	3.33	87.5	100	24.81
5	time for change, marcy winograd ...	0.76	50.0	100	13.70
38	Hansel Affair political ad	0.0	81.82	33.3	44.81
10	Mail Chimp Advertisement	16.67	29.59	12.5	18.93
24	Sonata, revamp, fuel, mileage, US	100	87.5	50	21.46
47	Vito Rocco Faintheart	100	57.14	0.0	40.62

6   Conclusions and Further Work

In this paper, we reported our investigation into the use of semi-supervised training for ASR on the Blip10000 corpus., including data segmentation, data selection, acoustic modelling and language modelling. We found that: (i) surprisingly equal segmentation was slight better than VAD segmentation of data due to its larger amount of useful training data, and (ii) that further gain in WERs can be obtained by adding 1-best transcripts of untranscribed data to n-gram and LSTM language model training data. We found that use of content genre embedding can add useful information of acoustic conditions to adapted acoustic model. Overall we achieve a 4% WER reduction on the dev set and 4.5% on the test set. However, these improvements increased SCR effectiveness by approximately 150% on the dev set and 77% on the test set.

Acknowledgement. This work was supported by Science Foundation Ireland as part of the ADAPT Centre (Grant 13/RC/2106) at Dublin City University.

References

1. Abdel-Hamid, O., Jiang, H.: Fast speaker adaptation of hybrid NN/HMM model for speech recognition based on discriminative learning of speaker code. In: Proceedings of ICASSP 2013, pp. 7942–7946 (2013)
2. Hadian, H., Sameti, H., Povey, D., Khudanpur, S.: End-to-end speech recognition using lattice-free mmi. In: Proceedings of Interspeech 2018, pp. 12–16 (2018)
3. Kudo, T., Richardson, J.: SentencePiece: a simple and language independent subword tokenizer and detokenizer for neural text processing. In: Conference on Empirical Methods in Natural Language Processing (EMNLP 2018), pp. 66–71 (2018)
4. Larson, M., Jones, G.J.F.: Spoken content retrieval: a survey of techniques and technologies. Found. Trends Inf. Retr. 4(4–5), 235–422 (2012)
5. Lüscher, C., et al.: RWTH ASR systems for librispeech: hybrid vs attention. In: Proceedings of Interspeech 2019, pp. 231–235 (2019)

6. Majumdar, S., Ginsburg, B.: MatchboxNet: 1D Time-channel separable convolutional neural network architecture for speech commands recognition. In: Proceedings of Interspeech 2020, pp. 3356–3360 (2020)
7. Manohar, V., Hadian, H., Povey, D., Khudanpur, S.: Semi-supervised training of acoustic models using lattice-free mmi. In: Proceedings of ICASSP 2018, pp. 4844–4848 (2018)
8. Narayanan, A., et al.: Toward domain-invariant speech recognition via large scale training. In: Proceedings of SLT 2018, pp. 441–447 (2018)
9. Panayotov, V., Chen, G., Povey, D., Khudanpur, S.: LibriSpeech: an ASR corpus based on public domain audio books. In: Proceedings of ICASSP 2015, pp. 5206–5210 (2015)
10. Povey, D., et al.: The Kaldi speech recognition toolkit. In: Proceedings of ASRU 2011, pp. 1–4 (2011)
11. Povey, D., et al.: Purely sequence-trained neural networks for ASR based on lattice-free mmi. In: Proceedings of Interspeech 2016, pp. 2751–2755 (2016)
12. Robertson, S.E., Walker, S., Jones, S., Hancock-Beaulieu, M., Gatford, M.: Okapi at TREC-3. In: Proceedings of TREC 3, vol. 500–225, pp. 109–126 (1994)
13. Sainath, T.N., et al.: A streaming on-device end-to-end model surpassing server-side conventional model quality and latency. In: Proceedings of ICASSP 2020, pp. 6059–6063 (2020)
14. Sanabria, R., et al.: How2: a large-scale dataset for multimodal language understanding. In: Proceedings of the Workshop on Visually Grounded Interaction and Language (ViGIL). NeurIPS (2018)
15. Saon, G., Soltau, H., Nahamoo, D., Picheny, M.: Speaker adaptation of neural network acoustic models using i-vectors. In: Proceedings of ASRU 2013, pp. 55–59 (2013)
16. Schmiedeke, S., et al.: Blip10000: a social video dataset containing SPUG content for tagging and retrieval. In: Proceedings of ACM MMSys 2013 (2013)
17. Snyder, D., Garcia-Romero, D., Sell, G., Povey, D., Khudanpur, S.: X-vectors: Robust DNN embeddings for speaker recognition. In: Proceedings of ICASSP 2018, pp. 5329–5333 (2018)
18. Stolcke, A.: SRILM-an extensible language modeling toolkit. In: Proceedings of International Conference on Spoken Language Processing (ICSLP 2002) (2002)
19. Veselý, K., Hannemann, M., Burget, L.: Semi-supervised training of deep neural networks. In: Proceedings of ASRU 2013, pp. 267–272 (2013)
20. Wang, Y., et al.: Espresso: a fast end-to-end neural speech recognition toolkit. In: Proceedings of ASRU 2019 (2019)
21. Xu, H., et al.: A pruned rnnlm lattice-rescoring algorithm for automatic speech recognition. In: Proceedings of ICASSP 2018, pp. 5929–5933 (2018)
22. Yu, K., Gales, M., Wang, L., Woodland, P.C.: Unsupervised training and directed manual transcription for LVCSR. Speech Commun. **52**(7), 652–663 (2010)

Various DNN-HMM Architectures Used in Acoustic Modeling with Single-Speaker and Single-Channel

Josef V. Psutka[1,2]([✉]) [iD], Jan Vaněk[2] [iD], and Aleš Pražák[2] [iD]

[1] Department of Cybernetics, University of West Bohemia, Pilsen, Czech Republic
[2] NTIS - New Technologies for the Information Society, UWB, Pilsen, Czech Republic
{psutka_j,vanekyj,aprazak}@kky.zcu.cz

Abstract. In this paper, we discuss some interesting features of training a special acoustic model for only one speaker with a constant acoustic background (acoustic channel). Currently, the LF-MMI method achieves the best results in many speech recognition tasks. A typical LF-MMI training procedure uses a special 1-state HMM topology that has different pdfs at the self-loop and forward transitions. We would like to discuss the replacement of this typical LF-MMI HMM by different types of HMM topologies (1-, 2- and 3-state HMM topologies that have outputs associated with states). Next, we discuss the advantages of using biphone context modeling over using the triphone context or even simpler context-free monophone. We also address the effect of the amount of training data and the context of DNN on WER, and all this with regard to a special acoustic model with one speaker and an almost constant acoustic channel.

Keywords: Speech recognition · Acoustic modeling · HMM topology · Lattice-free MMI · Single-speaker

1 Introduction

Lattice-free maximum mutual information (LF-MMI) DNN-HMM models [9] have achieved the state-of-the-art word error rates (WER) on various well-known speech databases such as Switchboard and Wall Street Journal in recent years [4,9]. However, all these databases have multiple speakers with different acoustic channels. In this paper, we would like to analyze the interesting properties of training the LF-MMI acoustic model (AM), specifically with respect to the single-speaker, single-channel task.

The conventional GMM-HMM topology in ASR has been a 3-state left-to-right HMM that can be traversed in at least 3 frames. This topology was replaced by a topology that can be traversed in one frame in a typical LF-MMI training procedure. This one-state HMM topology has different pdfs on the self-loop and forward transitions (observations are associated with arcs). This type of topology

© Springer Nature Switzerland AG 2021
L. Espinosa-Anke et al. (Eds.): SLSP 2021, LNAI 13062, pp. 85–96, 2021.
https://doi.org/10.1007/978-3-030-89579-2_8

is based on similarity to Connectionist Temporal Classification [3]. In this paper, we would like to analyze different types of HMM topologies.

In the article [9], a change of phonetic context modeling from triphone to left biphones in LF-MMI models was also proposed. We would like to discuss the possible advantages of using biphone context modeling over using the original triphone or the simpler monophone context on a single-speaker, single-channel task.

Last but not least, we would like to analyze the effect of the amount of training data on the quality of AM. And help to determine the minimum amount of data needed to train usable AM for this single-speaker, single-channel task.

The following section briefly describes the training and test data sets. In Sect. 3, we describe Acoustic feature extraction, different types of Acoustic modeling, Language Modeling, and Decoding. The experiments and results are described in Sect. 4. The conclusions are presented in Sect. 5.

2   Training and Test Data Set

All experiments were performed using a corpus that contained the speech of only one speaker. This corpus was created during training a shadow speaker. Shadow speakers are used in the respeaking (subtitling) of television programs. Thus, this is not a read-speech, rather we can consider this speech as spontaneous. For our experiments, we selected a speaker who spoke 146 h of speech. This amount of data was carefully annotated at the word level. No speech augmentation was performed (no additional noise or channel distortion was added). Thus, the corpus contains clean speech with a minimum of non-speech events. All data contain minimal background noise and were taken in the same acoustic environment over a relatively short period of time. The data were sampled at 16 kHz with 16-bit resolution.

The test part includes 3 h of speech by the same speaker. These 3 h consist of 18 different TV shows (10 min each) subtitled by this speaker. We chose the test data in this way to minimize the possible influence of the topic (and the associated language model) on recognition accuracy. From an acoustic point of view, there is no difference between the training and test conditions (see overview in Table 1)

Table 1. Training and test data set statistics.

	Train	Test
# of speakers	1	1
# of tokens	853k	13k
Dataset length [hours]	146	3

3 Experimental Setup

3.1 Acoustic Feature Extraction

The front-end is based on a Mel-scaled spectrum with 40 log energy filter-banks. These 40-dimensional features were used not only as input for DNN but also for GMM. Only in the case of training the baseline GMM-HMM model DCT, delta, and delta-delta features were added to the original coefficients. No normalization of the mean and variance was applied. Neither i-vectors nor other speaker adaptation techniques were used in the feature extraction process. All these methods are used for speaker adaptation, which makes no sense in the case of one speaker. Feature vectors were computed every 10 ms (100 frames per second) from 32 ms frames.

3.2 Acoustic Modeling

The structure and parameters of the acoustic models in the LVCSR system were tuned using KALDI toolkit [8].

GMM

All state-of-the-art neural networks used for acoustic modeling (with the exception of end-to-end approaches) need information about the alignment of some typically subword units for their training. These sub-word units are monophones or clustered states of triphones or biphones [6,9]. It was analyzed in [5] that the lower the error rate of the GMM-HMM system used in the forced alignment to generate the frame-level training labels for the neural network, the lower the error rate of the resulting neural network-based system. This effect was consistent across all types of DNNs.

The elementary speech unit in our GMM-HMM was represented by a 3-state HMM with a continuous fully covariance probability density function assigned to each state. Several experiments were performed to find the optimal number of clustered states of triphones (the optimal number of states was 3383). Also, the number of mixtures of multivariate Gaussians was optimized according to the paper [11] and [10] (to a likelihood accuracy of $\beta \geq 0.7$). No feature-space Maximum Likelihood Linear Regression (fMLLR) or Speaker-Adaptive Training procedure (SAT) was used to fit the GMM models, as we only have one speaker and one channel.

TDNN_CE

Time Delay Neural Networks (TDNN) with cross-entropy (CE) training criterion have shown to be effective in modeling long-range temporal dependencies [15]. The TDNN_CE training procedure was slightly modified compared to those presented in [6]. The first splicing was the Linear Discriminant Analysis (LDA) transforms layer $(-2, -1, 0, 1, 2)$. Subsequent layers then had contexts $(-1, 0, 1)$, $(-1, 0, 1)$, $(-3, 0, 3)$ and $(-6, -3, 0)$. The $(-1, 0, 1)$ means that the first layer sees 3 consecutive frames of input thus the $(-3, 0, 3)$ means that the hidden layer sees 3 frames of the previous layer, separated by 3 frames. In total, we have five hidden layers with the ReLu activation function with 650 nodes and a final softmax output layer, which computes posteriors for output states (number of GMM-derived states was 3383). The overall context is therefore 13 frames to the left and 7 to the right. The state-level Minimum Bayes Risk (sMBR) [14] has been used to improve recognition accuracy.

TDNNF_LF-MMI
Povey et al. [9] applied Maximum Mutual Information (MMI) training with DNN-HMM models using a full denominator graph (hence the name Lattice-Free) by using a phone language model (instead of a word language model). Instead of a frame-level objective, the log-probability of the correct phone sequence as the objective function is used. The LF-MMI (Lattice-Free Maximum Mutual Information) training procedure has a sequence discriminative training criterion without the need for frame-level cross-entropy pre-training. In regular LF-MMI, all utterances are split into fixed-size chunks (usually 150 frames) to make GPU computations efficient [9]. A typical LF-MMI topology has 12 TDNNF layers (factorized TDNN [7]), a hidden layer dimension of 1024, a bottleneck dimension of 128, and a context of ± 29, i.e., a context per layer is (1 1 1 1 3 3 3 3 3 3 3).

CNN-TDNNF_LF-MMI
It has been shown [1] that locality, weight sharing, and convolutional layer pooling have the potential to improve ASR recognition accuracy. The typical Kaldi CNN-TDNN models consist of 6 CNN layers followed by 10 TDNNF (factorized TDNN [7]) layers and two output layers: chain based (LF-MMI criterion) and cross-entropy criterion (xent). The first convolutional layer receives at input three vectors of speech features (the current, previous, and next acoustic frames). It uses 64 filters of size 3×3 to perform time and feature space convolutions and outputs a $64 \times 40 \times 1$ volume. The following convolutional layers apply more filters (128 and finally 256), but preserve the size of the feature volume by decreasing the height from 40 to 20 and finally to 10.

3.3 Language Modeling

We used two types of language models in our experiments. The first was a trigram language model (LM) with mixed-case vocabularies of more than 1.2 million words. Our training text corpus contains data from newspapers (520M tokens), web news (350M tokens), subtitles (200M tokens), and transcripts of some TV shows (175M tokens) (details can be found in [12]). The resulting LM has 35M bigrams and almost 30M trigrams. The percentage of OOV words in the test set is 0.44%.

In order to better analyze possible improvements to the acoustic model (and minimize the influence of the language model), we used a simplified version of the previously described language model. This simplification consists of using only the unigram probabilities of the language model.

3.4   Decoding

All recognition experiments were performed using our in house real-time ASR system. This LVCSR system is optimized for low latency in the real-time (RT) operation with very large vocabularies. To enable recognition with a dictionary containing more than one million words in RT, we accelerated decoding using a parallel approach (Viterbi search on the CPU and DNN segment scores on the GPU [13]).

4   Experiments

4.1   HMM-topology and Context-Dependent Modeling

The conventional three-state left-to-right GMM-HMM topology was replaced by a special single-state topology in a typical LF-MMI training procedure (see $1state_{arc}$ i.e. b) in Fig. 1). This topology has different pdfs on the self-loop and forward transitions (highlighted in bold in the figure). In this case, observations are associated with arcs (rather than states). All HMM transition probabilities are fixed and uniform because training and applying them does not improve the recognition results (see [9]).

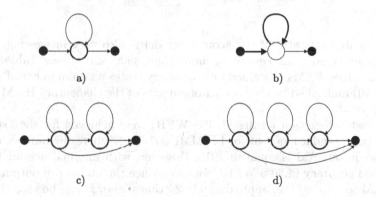

Fig. 1. Different HMM topologies. The circles represents the states and arrows the transitions. Input and output states are non-emitting and are shown in black. a) $1state$, b) $1state_{arc}$, c) $2state\_skip$ and d) $3state\_skip$

As previous studies [2] have shown, context-dependent (CD) modeling is an essential component of HMM-based models, not only for GMM-HMM but also for DNN-HMM. Commonly used triphones (in GMM-HMM based models) have been replaced by biphones in the currently used DNN-HMM approaches. This change led not only to a significant reduction in the number of states of the acoustic model but also to a simplification of the decoding since by using the left biphone, the problems with the right context in the transition between words are eliminated during decoding.

An alternative to CD modeling is context-independent (CI) modeling, i.e., monophone modeling. CI models have a much smaller number of states. This number is equal to the number of phonemes multiplied by the number of emitting states of the elementary HMM model. When using the CI approach, the context problem is then completely eliminated during decoding.

In Table 2 we can see the effect of different HMM topologies and phonetic context-dependent modeling on the resulting WER in the TDNN LF-MMI approach. In this case, we used a typical Kaldi setup for this type of DNN (i.e., 12 TDNNF layers; the hidden layer dimension is 1024, the bottleneck dimension is 128, and the context is ± 29). Note that all recognition results were obtained using a unigram language model to make the differences between variants more obvious.

Table 2. Impact of HMM topology and context-dependent modeling on recognition results in the TDNN LF-MMI approach.

	WER %		
	Monophone	Biphone	Triphone
$1state$	5.21	4.72	5.03
$1state_{arc}$	4.80	4.70	5.04
$2state\_skip$	5.07	4.70	5.02
$3state\_skip$	5.00	4.78	5.08

The number of states of the acoustic model is also very interesting. While the triphone-based AMs contained more than 3300 states (see Table 3), the monophone-based AMs contained only as many states as the number of phonemes (i.e. 40) multiplied by the number of outputs of the elementary HMM model (i.e. 1, 2, or 3).

The best recognition results (4.70% WER) were achieved for the $1state_{arc}$ biphone (the standard variant for LF-MMI training), where the number of output states in our AM is equal to 1912. However, with minimal degradation in recognition accuracy (4.80% WER), we can reduce the number of output states in the AM to only 80 (i.e. approximately 24 times less). It can be seen that the biphone AMs perform slightly better than monophone or triphone HMM models. Let us also note that for both biphones and triphones AMs, the optimal number of states was sought to find the best recognition results.

Table 3. Number of AM output states depending on the phonetic context and the topology of the elementary HMM model.

	Monophone	Biphone	Triphone
$1state$	40	992	3448
$1state_{arc}$	80	1912	3400
$2state\_skip$	80	1824	3352
$3state\_skip$	120	1904	3344

Similar experiments were also performed for the CNN-TDNN LF-MMI approach. For these experiments, we used the default topology described in Sect. 3.2. From the obtained results (see Table 4), it is evident that no improvement (or only marginal improvement for the monophone based model) was achieved by using convolutional layers.

Using convolutional layers to handle some typical speech variables and variability, e.g., vocal tract length differences across speakers, distinct speaking styles causing formant undershoot or overshoot, etc. (explicitly expressed in the frequency domain) does not yield any significant improvement in the single-speaker, single-channel task.

Table 4. Impact of HMM topology and context-dependent modeling on recognition results in the CNN-TDNN LF-MMI approach.

	WER %		
	Monophone	Biphone	Triphone
$1state$	5.17	4.89	5.02
$1state_{arc}$	4.81	4.77	5.10
$2state\_skip$	4.98	4.83	5.17
$3state\_skip$	4.92	4.81	5.04

4.2   DNN Context

The size of the DNN context and the associated AM delay is one of the important issues in optimizing a speech recognition system for real-time use. In this section we analyze the effect of the DNN context on the recognition accuracy, not only for the conventional acoustic model setup described in Sect. 3.2, but also for selected prospective variants of the LF-MMI approaches, see Sect. 4.1 (i.e. $1state$ monophone, $1state_{arc}$ monophone and $1state_{arc}$ biphone).

We varied the context from ±5 to ±47, but only for the TDNN LF-MMI variant without convolutional layers (the CNN layers did not yield any significant improvement for the single-speaker, single-channel task). The results are listed in Table 5 and shown in Fig. 2.

4.3   Implementation Issues

Unfortunately, our LVCSR system is based on the classical HMM [17] implementation. "HTK-style" HMM observations are associated with states, not arcs. But both approaches can be transformed to each other. Figure 3 shows an example of such a transformation. As you can see, both HMMs can emit the same outputs: either a, or ab, or abb, etc. However, the figure shows that after the transformation, the original 1-state *arc*-based model (with different pdfs associated with self-loop and forward transitions) has changed to a 2-state *state*-based HMM.

Table 5. DNN context for selected HMM topologies.

| | WER % | | |
| | Monophone | | Biphone |
Context ±	$1state$	$1state_{arc}$	$1state_{arc}$
5	6.28	5.71	5.44
8	5.74	5.49	5.11
11	5.46	5.20	4.98
14	5.37	5.10	4.94
17	5.36	4.99	4.82
20	5.34	4.95	4.85
23	5.20	4.90	4.79
26	5.20	4.88	4.71
29	5.21	**4.80**	**4.70**
32	**5.17**	4.85	4.76
35	5.19	4.89	4.72
38	5.18	4.86	4.74
41	5.18	4.83	4.76
44	5.17	4.81	4.74
47	5.18	4.82	4.72

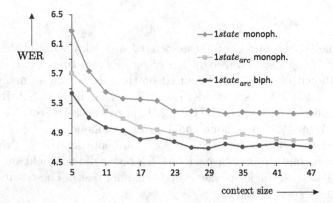

Fig. 2. Dependence of WER on DNN context for the LF-MMI training procedure for selected HMM topologies.

Therefore, also applying the arc-based model to monophones, for example, has twice the number of outputs as the classical 1-state $state$-based model. For this reason, the 1-state arc-based model can be considered more like a 2-state model (in the classical sense, where outputs are associated with $states$).

Fig. 3. An example of *arc*-based transformation to *state*-based HMM.

4.4 Amount of Training Data

How much data do we need to train a good acoustic model? This question is very often discussed for general speaker-independent acoustic models with different kinds of acoustic backgrounds. Recall that all our data are spoken by a single speaker in the same acoustic environment in a relatively short period of time. These data can be considered constant from an acoustic modeling point of view. However, for this task, this question has not been answered in a relevant way yet.

Table 6. Amount of training data.

# of training hours	10		30		50		70		90		110		130		146
		20		40		60		80		100		120		140	
% WER	6,37		5,42		5,04		5,05		4,91		4,75		4,68		4,70
		6,04		5,10		5,05		5,02		4,88		4,67		4,71	

We randomly divided the entire training data into 10-hour chunks and gradually combined them to create larger training sets. Thus, the smallest segment corresponded to only 10 h and the longest to the entire training set, i.e. 146 h. The dependence of WER on the amount of training data is listed in Table 6 and shown in Fig. 4.

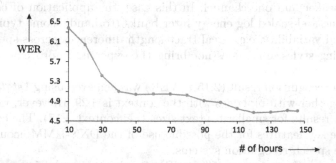

Fig. 4. Dependence of WER on the amount of training data.

Table 7. Summary of the recognition results

		WER [%]	
		Unigram LM	Trigram LM
GMM		7.81	3.81
TDNN_CE		7.64	3.69
	sMBR	6.64	2.79
TDNN_LF-MMI	$1state$ monoph.	5.17	2.45
	$1state_{arc}$ monoph.	4.80	2.20
	$1state_{arc}$ biph.	**4.70**	**2.15**
CNN-TDNNF_LF-MMI	$1state$ monoph.	5.17	2.40
	$1state_{arc}$ monoph.	4.81	2.23
	$1state_{arc}$ biph.	4.77	2.22

5   Conclusion

This paper discussed and compared some interesting parameters of acoustic modeling using TDNN LF-MMI for special single-speaker and single-channel task. The influence of HMM topology was investigated, 1- 2- and 3-states HMM were tested and compared. Various context dependencies (monophone, left biphone and triphone) were also investigated. The effect of adding convolutional layers to the TDNN topology was tested. The effect of context width on recognition accuracy for selected perspective AM configurations (depending on the phonetic context and the topology of the basic HMM model) was also investigated. Finally, the amount of data required to train the acoustic model was also analyzed. The most interesting results are summarized in Table 7 (recall that the OOV is 0.44%).

It is clear from the results that the use of CNN TDNN_LF-MMI does not yield any significant recognition improvement for this special single-speaker, single-channel task. This is probably due to the specific acoustic setup where there is only one speaker and one channel. In this case, the application of convolution kernels to the Mel-scaled log energy filter-banks (to handle some typical speech variables and variability, e.g. vocal tract length differences across speakers, distinct speaking styles, etc.) does not bring the expected improvement that has been common in other tasks.

The best recognition result (2.15% WER) was achieved using $1state_{arc}$ HMM topology together with biphones with the context is ±29. However, we obtained very similar results for smaller context sizes (i.e. context ±17). The context size is one of the key features for the possible use of the DNN-HMM acoustic model in real-time speech recognition systems.

A very interesting result (2.45% WER) was also achieved using a monophone model with a $1state$ HMM topology. In this case, there were as many DNN outputs as there were different phonemes in the task (i.e. 40). Although there was a deterioration of 0.3% (compared to the best result of $1state_{arc}$ with biphones), this result was very promising because using monophones instead of biphones could significantly simplify the whole decoding process.

The amount of training data needed to train a good acoustic model for single-speaker and single-channel was estimated to be at least 120 h.

In the future, we would like to perform the same analysis on a larger number of single-speakers single-channel tasks to verify the generality of the obtained results. We would also like to analyze the influence of the DNN context, especially the asymmetric context. In particular, we are interested in its possible minimization in the future with respect to the results of recognition and use in real-time.

We have also tried the very promising end-to-end architecture [16]. Although this task seems ideal for their use in some aspects, in our first experiments we did not achieve better recognition results than using current state-of-the-art conventional DNN systems. The application of new end-to-end approaches will also be the subject of our future work.

Acknowledgments. This research was supported by the ITI project of the Ministry of Education of the Czech Republic CZ.02.1.01/0.0/0.0/17_048/0007267 InteCom.

References

1. Abdel-Hamid, O., Mohamed, A., Jiang, H., Deng, L., Penn, G., Yu, D.: Convolutional neural networks for speech recognition. IEEE/ACM Trans. Audio Speech Lang. Process. **22**(10), 1533–1545 (2014)
2. Deng, L., Acero, A., Dahl, G., Yu, D.: Context-dependent pre-trained deep neural networks for large vocabulary speech recognition. IEEE Trans. Audio Speech Lang. Process. **20**, 30–42 (2012)
3. Graves, A., Fernández, S., Gomez, F.: Connectionist temporal classification: labelling unsegmented sequence data with recurrent neural networks. In: In Proceedings of the International Conference on Machine Learning, ICML 2006, pp. 369–376 (2006)
4. Han, K.J., Hahm, S., Kim, B., Kim, J., Lane, I.R.: Deep learning-based telephony speech recognition in the wild. In: Interspeech 2017, pp. 1323–1327 (2017)
5. Hinton, G., et al.: Deep neural networks for acoustic modeling in speech recognition: the shared views of four research groups. IEEE Signal Process. Mag. **29**(6), 82–97 (2012)
6. Peddinti, V., Povey, D., Khudanpur, S.: A time delay neural network architecture for efficient modeling of long temporal contexts. In: Interspeech 2015, pp. 3214–3218 (2015)
7. Povey, D., et al.: Semi-orthogonal low-rank matrix factorization for deep neural networks. In: Interspeech 2018, pp. 3743–3747 (2018)
8. Povey, D., et al.: The Kaldi speech recognition toolkit. IEEE 2011 Workshop on Automatic Speech Recognition and Understanding (2011)
9. Povey, D., et al.: Purely sequence-trained neural networks for ASR based on lattice-free MMI. In: Interspeech 2016, pp. 2751–2755 (2016)

10. Psutka, J.V.: Gaussian mixture model selection using multiple random subsampling with initialization. In: Azzopardi, G., Petkov, N. (eds.) CAIP 2015. LNCS, vol. 9256, pp. 678–689. Springer, Cham (2015). https://doi.org/10.1007/978-3-319-23192-1_57
11. Psutka, J.V., Psutka, J.: Sample size for maximum-likelihood estimates of gaussian model depending on dimensionality of pattern space. Pattern Recogn. **91**, 25–33 (2019)
12. Švec, J., Hoidekr, J., Soutner, D., Vavruška, J.: Web text data mining for building large scale language modelling corpus. In: Habernal, I., Matoušek, V. (eds.) TSD 2011. LNCS (LNAI), vol. 6836, pp. 356–363. Springer, Heidelberg (2011). https://doi.org/10.1007/978-3-642-23538-2_45
13. Vaněk, J., Trmal, J., Psutka, J.V., Psutka, J.: Optimized acoustic likelihoods computation for NVIDIA and ATI/AMD graphics processors. IEEE Trans. Audio Speech Lang. Process. **20**(6), 1818–1828 (2012)
14. Veselý, K., Ghoshal, A., Burget, L., Povey, D.: Sequence-discriminative training of deep neural networks. In: Interspeech 2013, pp. 2345–2349 (2013)
15. Waibel, A., Hanazawa, T., Hinton, G., Shikano, K., Lang, K.J.: Phoneme recognition using time-delay neural networks. IEEE Trans. Acoust. Speech Signal Process. **37**(3), 328–339 (1989)
16. Wang, D., Wang, X., Lv, S.: An overview of end-to-end automatic speech recognition. Symmetry **11**(8), 1018 (2019)
17. Young, S.: The HTK hidden Markov model toolkit: design and philosophy. Entrop. Camb. Res. Lab. Ltd. **2**, 2–44 (1994)

Invariant Representation Learning for Robust Far-Field Speaker Recognition

Aviad Shtrosberg[1](✉), Jesus Villalba[2](✉), Najim Dehak[2](✉), Azaria Cohen[1], and Bar Ben-Yair[3]

[1] The Open University of Israel, Raanana, Israel
aviad.shtrosberg@mail.huji.ac.il
[2] Center for Language and Speech Processing, Johns Hopkins University, Baltimore, MD, USA
{jvillalba,ndehak3}@jhu.edu
[3] Penta-AI, Tel-Aviv, Israel
bbenyai1@jhu.edu

Abstract. The emergence of smart home assistants increased the need for robust Far-Field Speaker Identification models. Speaker Identification enables the assistants to perform personalized tasks. Smart home assistants face very challenging speech conditions, including various room shapes and sizes, various distances of the speaker from the microphone, various types of distractor noises (TV in the background, air conditioner, fridge, babble speech of other speakers, etc.). This paper describes the use of Invariant Representation Learning (IRL) as a method aimed to increase the robustness of Speaker Identification models on Far-Field. We introduce three new versions of IRL: Text-Dependent IRL (TD-IRL), Text Independent IRL (TI-IRL), and Deep Features IRL (DF-IRL). We evaluate the IRL models performance and compare them to the base x-vector model. The various Far-Field scenarios are evaluated using VOiCES dataset - a dataset of simulated Far-Field recordings in four real furnished rooms. TD-IRL and DF-IRL improve the minDCF results on the far-field scenarios by an average of 36%, and TI-IRL improves it by 31% with respect to the baseline model.

Keywords: Speaker identification · Far-field · Invariant Representation Learning (IRL) · Text-Dependent-IRL (TD-IRL) · Text-Independent-IRL (TI-IRL) · Deep-Features-IRL (DF-IRL)

1 Introduction

Speech technologies are becoming more prevalent in our daily lives and have become one of the primary means for interacting with technologies, including smart home assistants, smartphones, smart-watches, spoken commands to cars, spoken search engines queries, etc. All of those technologies need to work in real-life scenarios, which include all sorts of background noises. This paper will focus on improving the performance of automatic speaker verification (ASV) in

© Springer Nature Switzerland AG 2021
L. Espinosa-Anke et al. (Eds.): SLSP 2021, LNAI 13062, pp. 97–110, 2021.
https://doi.org/10.1007/978-3-030-89579-2_9

far-field scenarios for smart home assistants. Speaker recordings in noisy environments sound different from related "clean" recordings. To overcome this problem, most speech and speaker recognition systems are trained using augmented noise to increase their robustness [11,18]. In recent years, the state-of-the-art method for speaker verification is based on the x-vector approach [19]. However, the x-vector models do not perform optimally in far-field scenarios yet. This work aims to introduce changes to the x-vector network architecture and training method to improve the model's robustness to far-field. The original x-vector architecture was based on Time Delay Neural Network (TDNN), and the classification layer was a simple softmax layer. Since then, some modifications were made to the original architecture. In [6], the TDNN encoder was replaced by a ResNet34 encoder network. In [12], the Cross-Entropy loss over the Softmax layer was replaced by Angular-Softmax (A-Softmax) loss. In [3], Additive Margin Softmax (AM-Softmax) took its place, since it converged better at training.

The x-vector network training process includes data augmentation with MUSAN [18] for adding noise and room impulse responses (RIR) [11] for adding reverberations to the utterances. The augmentation process is used for increasing the robustness of the network to different types of background noises. Another methods for increasing the model's robustness to noise include speech enhancement [9,10] dereverberation and beamforming [14]. In [13] Invariant representation learning (IRL) was shown to improve speech recognition acoustic model over the standard data augmentation. IRL is a method for training the network over clean and noisy features of the same utterance that forces the network to generate close embeddings from the clean and noisy features by adding a loss over the distance between their embeddings. In [7] IRL was adopted to the task of speaker verification.

This paper proposes and evaluates three variants of invariant representation learning to train x-vector architectures robust to far-field. The base x-vector model used for the evaluation is from [20]–The front-end x-vector is ResNet34 with AM-Softmax plus PLDA [8] back-end.

The new methods are called: Text-Dependent IRL (TD-IRL), Text Independent IRL (TI-IRL), Deep Features IRL (DF-IRL). The Deep Features version was inspired by [10]. We evaluate each model's robustness to different Far-Field scenarios using VOiCES dataset [17]. In this paper, enrollment is done on "clean" utterances, and the test is on Far-Field utterances. The original IRL version is text-dependent, meaning that it is trained on clean-noisy pairs, which need to be simulated or recorded simultaneously through different devices. This paper introduces a Text-Independent version of IRL, which removes the "same-text" requirement in the original IRL method. This enables us to use recordings of the same speaker from various sources. In this paper, we experiment with combining the deep feature (DF) loss speech enhancement method [10] with the IRL method in the DF-IRL model. We evaluate each model's robustness to the different Far-Field scenarios. In Sect. 2, we will describe the Base Model architecture. In Sect. 3, we describe Invariant Representation Learning (IRL). Section 4 describes the experimental setup. In Sect. 5, we display the results, and Sect. 6 contains the conclusions from the evaluation.

2   Base Model Architecture

All the IRL models were finetuned from the base model and based on its architecture [20]. We finetuned the models for 35 epochs with a learning rate of 0.05 and additive margin warm-up of 20 epochs with frozen pre-embedding weights (only the output layer and the last affine transformation before the embedding are updated). We used a learning rate scheduler that exponentially decays, divided by 2 every 8,000 model updates. The base model used in the paper is x-vector with ResNet34 encoder network and AM-Softmax loss. The input of the encoder network is log-filterbanks. Features were short-time mean normalized. We applied energy-based VAD for filtering the non-speech frames.

2.1   Encoder Network: ResNet34

Table 1 summarizes the topology of the ResNet34 encoder. There are four types of residual blocks with different number of channels in the 2D convolutions. Each type of block is repeated multiple times, as indicated in the Blocks column of the table. Each time we increase the number of convolution channels, we downsample the filter-bank and time dimensions by 2. Finally, we flatten channel and frequency dimensions to get a single vector per frame.

Table 1. ResNet34 encoder architecture for 8 kHz systems

Layer	Structure	Output
Input	–	$64 \times 1 \times T$
Conv2D-1	$3 \times 3, 64$, Stride 1	$64 \times 64 \times T$
ResNetBlock-1	$\begin{bmatrix} 3 \times 3, 64 \\ 3 \times 3, 64 \end{bmatrix} \times 3$, Stride 1	$64 \times 64 \times T$
ResNetBlock-2	$\begin{bmatrix} 3 \times 3, 128 \\ 3 \times 3, 128 \end{bmatrix} \times 4$, Stride 2	$32 \times 128 \times T/2$
ResNetBlock-3	$\begin{bmatrix} 3 \times 3, 256 \\ 3 \times 3, 256 \end{bmatrix} \times 6$, Stride 2	$16 \times 256 \times T/4$
ResNetBlock-4	$\begin{bmatrix} 3 \times 3, 512 \\ 3 \times 3, 512 \end{bmatrix} \times 3$, Stride 2	$8 \times 512 \times T/8$
Flatten	–	$4096 \times T/8$
StatsPooling	–	8192
Dense	–	256

2.2   Classification Layer: Additive Margin Softmax (AM-Softmax)

The loss function used by the base model is called AM-Softmax [3,21]. This loss function aims to increase the between-class variance of the embedding layer while decreasing the within-class variance. Thus, it makes the x-vectors of a speaker more separable from the vectors of other speakers. The margin of the function is introduced in an additive manner, starting at a margin equals to zero and increasing it at each training epoch.

This loss function is based on the cross-entropy loss. In the following equation, Cross-Entropy is written as a function of the weights of the output layer (W) and the activations at the previous layer (f_i)

$$L_{\text{CE}} = -\frac{1}{n} \sum_{i=1}^{n} \log \frac{e^{W_{y_i}^T f_i}}{\sum_{j=1}^{C} e^{W_j^T f_i}}, \tag{1}$$

where i is the sample index and y_i is the label index for sample i. By expressing inner products as $u \cdot v = ||u|| \cdot ||v|| \cos(\theta)$ where θ is the angle between the vectors, we get the angular form of the cross entropy loss:

$$L_{\text{CE}} = -\frac{1}{n} \sum_{i=1}^{n} \log \frac{e^{||W_{y_i}|| \cdot ||f_i|| \cos(\theta_{y_i})}}{\sum_{j=1}^{C} e^{||W_j|| \cdot ||f_i|| \cos(\theta_j)}}. \tag{2}$$

By normalizing the norms of the weight vector and the previous layer output (making $||W_i||, ||f||$ to be 1) and by generalizing the target logit from $\cos(\theta)$ to $\psi(\theta) = \cos(\theta) - m$ we get AM-Softmax loss function:

$$L_{\text{AMS}} = -\frac{1}{n} \sum_{i=1}^{n} \log \frac{e^{s \cdot (\cos \theta_{y_i} - m)}}{e^{s \cdot (\cos \theta_{y_i} - m)} + \sum_{j=1, j \neq y_i}^{C} e^{s \cdot \cos \theta_j}} \tag{3}$$

Where m is a positive integer that controls the angular margin size and s is a hyperparameter that scales the cosine values. The normalization step on features and weights makes the predictions only depend on the angle between the feature and the weight. The learned embedding features are thus distributed on a hypersphere with a radius of s. As the embedding features are distributed around each feature center on the hypersphere, we add an additive angular margin penalty m between x_i and W_{y_i} to simultaneously enhance the intra-class compactness and inter-class variance.

2.3   Back-End

All the models described in this paper used the same probabilistic linear analysis (PLDA) [8] based back-end to compare the enrollment and test x-vectors. x-vectors were preprocessed using the standard pipeline. First, we applied linear discriminant analysis (LDA) for reducing the 256 dimensions embedding to 150. Then, we applied centering and whitening, and finally, we projected the x-vectors to the unit hypersphere by applying length normalization. Finally, Gaussian Simplified PLDA (SPLDA) was used to compute the log-likelihood ratio between the target and non-target hypotheses.

3   Invariant Representation Learning (IRL)

Invariant Representation Learning is a method that was shown to improve speech recognition acoustic model [13] over the standard data augmentation. In [7], the method has been adapted for training x-vector embedding. The training method of IRL, as suggested in [7], is as follows. At each training iteration, a pair of clean x and noisy x' versions of the same utterances ("same text") is fed into the network one by one, then three types of losses are calculated:

1. $L_C(x), L_C(x')$ - Categorical Loss for the clean and noisy utterances respectively
2. $L_{cos}(x, x')$ - Cosine Distance between the embedding vectors of the clean and noisy utterances
3. $L_2(x, x')$ - Euclidean loss between the embedding vectors of the clean and noisy utterances

Those losses are combined and represented by:

$$L_{IRL}(x, x') = L_C(x) + \alpha L_C(x') + \gamma L_{cos}(x, x') + \lambda L_2(x, x') \tag{4}$$

where α, γ, λ are hyperparameters that control the weight of each loss in the IRL loss. The addition of the Cosine and Euclidean losses leads the network to map the clean and noisy utterance embeddings to a similar space.

In this paper, we modify this basic version of IRL and introduce three additional methods inspired by it:

3.1   Text-Dependent IRL (TD-IRL)

This version is similar to the original IRL (from [7]) but removing the L_2 loss. This is because when vectors are normalized to unit length, euclidean distance is equivalent to cosine distance. We assigned equal weights to the Categorical Losses of the clean and noisy utterances and weight $\gamma = 0.01$ to the Cosine Distance loss. We call this version of IRL: Text-Dependent since we use only pairs of clean and noisy versions of the same utterance where the noisy version is a simulated Far-Field recording. In this method, the compactness of the embedding distribution for each speaker is achieved through AM-Softmax. And the IRL makes each noisy utterance embedding close (in terms of Cosine Distance) to its clean counterpart. Therefore, clean and noisy utterances fall under the same Gaussian modeled by the PLDA. IRL is related to NAP and PLDA by the essence that both techniques try to make the embedding invariant to noise; but IRL takes advantage of the non-linear behavior of the neural network.

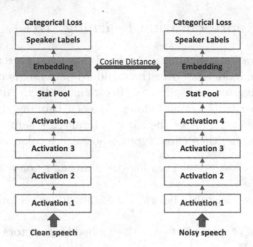

Fig. 1. Visualization of the training process used both at TD-IRL and TI-IRL

3.2   Text Independent IRL (TI-IRL)

This version of IRL uses the same loss function as Text-Dependent-IRL. Here we changed the data loader method. Instead of loading pairs from the same utterance in clean and noisy versions, in this IRL version, at each training itera-tion, we take two utterances with "different-text" (clean and noisy) of the same speaker. In this method, we force the cluster of all the noisy embeddings from each speaker to be close to the cluster of embeddings from its clean utterances. This means that the compactness of the clean and noisy vectors is enforced not only by AM-Softmax but also by the Text Independent IRL. This version of the model does not limit us to use "same-text" pairs of (clean, noisy) utterances. Creating the "same-text" pairs can only be done in a recording studio setup when simulating the Far-Field recordings or noise augmentation. The text-independent pairs can be sampled from any recording platform as long as both the clean and noisy recordings belong to the same speaker. Thus, TI-IRL is a more flexible training method allowing us to use real data without resorting to simulation or complicated audio capture setups.

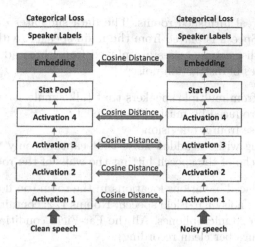

Fig. 2. DF-IRL training process visualization

3.3 Deep Features IRL (DF-IRL)

This version of IRL takes only pairs of the same utterance (like TD-IRL). It was inspired by, [10] which proposes the use of loss over the deep activation layers of the x-vector network for training a speech enhancement network. Here we add the Cosine Distance between all the activation layers to the loss function $L_{IRL}(x, x')$, namely, the loss is calculated over conv1, resblock-1x, resblock-2x, resblock-3x, resblock-4x (see Table 1). The training process is visualized in Figs. 1 and 2. We assign equal weight to all the activation losses. The final loss is:

$$L_{\mathrm{DF-IRL}}(x, x') = L_C(x) + \alpha L_C(x') + \gamma L_{\cos}(x, x') + \lambda L_{\mathrm{activations}}(x, x') \quad (5)$$

where:

$$L_{\mathrm{activations}}(x, x') = \sum_{a_i \in \mathrm{ActivationLayers}} \delta_i \cdot L_{\cos}(a_i(x), a_i(x')) ; \quad (6)$$

δ_i is the weight of the loss over the activation layers, and $a_i()$ is the output of the activation layer i. We assigned equal weights to γ, λ in our evaluation of this model. Adding the loss over the activation layers aims to force the deeper layers of the network to ignore the noise. That incorporates speech enhancement to the deep layers of the network.

4 Experimental Setup

4.1 VOiCES Dataset

Voices Obscured in Complex Environmental Settings (VOiCES) dataset [17] consists of 4000 h of human speech from 300 different speakers. The clean recordings were taken from the LibriSpeech Corpus [16]. The Far-Field recordings were simulated in four real furnished rooms. The data simulates different Far-Field conditions. Clean Speech is played from the main speaker (with simulated head movements), and distraction noises are played from three additional speakers. There are four types of distraction noises:

- Babble – noise from multiple speakers that talk simultaneously
- Music – noise produced by musical instruments
- Television – noise from a television
- None – recording without added noise that captures only the reverberation caused by the echo of the speech hitting the walls of the room.

Microphones are placed in various locations in the room (on floor, table, ceiling, wall) and record the noisy Far-Field speech. Rooms 1, 2: contain 12 microphones; rooms 3, 4: contain 20 microphones. All the Far-Field conditions are annotated (256 Noisy recordings per clean recording).

4.2 Base Model Training

The base model was trained on telephony speech with 8 KHz sample rate, augmented with MUSAN and RIR. The reason we used 8 KHz instead of 16 KHz recordings is to make the model compatible with the lower computing resources found in personal assistant devices.

The datasets used for training:

- Mixer6 [1] - telephone and microphone speech
- SRE [4] Telephony: SRE04-06 telephone speech, SRE08, SRE08 Supplemental, SRE10 Tel , SRE10 interview and mic phone calls, SRE12
- Voxceleb 1+2 [2,15]
- SWBD (Switchboard) [5].

4.3 Data Split for Evaluation

Train Set: We finetuned the base models on the VOiCES Train Set rooms 1, 2, 3 only. As a baseline, we finetuned the models using only AM-softmax loss on clean and noisy utterances sampled randomly from the training set. For the IRL models, we created a data loader that randomly samples (clean, noisy) pairs. For TD-IRL and DF-IRL, the data loader samples (clean, noisy) pairs of the same speaker with "same text" and for the TI-IRL, it samples (clean, noisy) pairs of the same speaker with "different text".

Enrollment: We used the clean speech of the speakers from VOiCES Test Set (No overlap between train and test speakers). Enrollment was done on the first 5 utterances of the speaker.

Test: We used the noisy utterances from VOiCES Test Set only from room 4. We split to different Far-Field test scenarios by the microphone number. Each microphone is located in a different location in the room with a different distance to the main speaker and all the distractor loudspeakers. Therefore it represents a different Far-Field scenario. At each Far-Field scenario, we included all types of distractor noises (Babble, Music, Television, None). We excluded the first 5 utterances and all its noisy counterparts of the speaker from the test set (to prevent overlap between enrollment and test data). As opposed to other existing methods that use microphone arrays and beam forming techniques for speech enhancement, our models use recordings from a single microphone at each test condition.

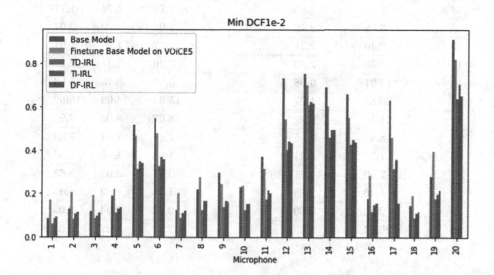

Fig. 3. Plot of minDCF1e-2 result per microphone

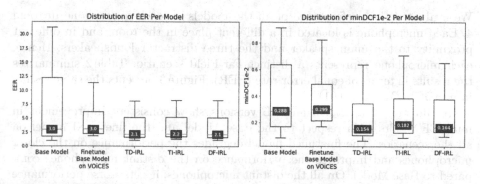

Fig. 4. EER and minDCF1e-2 distribution per model all room 4 microphones

Table 2. EER of all the models per microphone in room 4

Microphone	Base model	Finetune base model on VOiCES	TD-IRL	TI-IRL	DF-IRL
1	0.93	1.95	1.06	1.25	1.11
2	1.2	1.83	1.3	1.47	1.3
3	1.07	2.09	1.25	1.5	1.35
4	1.82	2.26	1.62	1.73	1.63
5	10.27	5.88	4.62	4.88	4.76
6	10.96	6.16	4.92	5.04	4.99
7	1.35	2.03	1.3	1.46	1.36
8	2.15	2.55	1.73	1.74	1.71
9	3.09	2.48	2.06	2.09	2.07
10	2.46	2.31	1.87	1.96	1.91
11	3.79	2.83	2.26	2.4	2.22
12	15.91	6.08	5.67	5.75	5.63
13	30.83	11.27	15.24	14.99	15.62
14	21.16	8.2	8.07	8.06	8.2
15	18.4	7.95	7.67	7.84	7.89
16	1.77	3.21	1.64	1.86	1.64
17	7.69	4.41	3.71	3.8	3.63
18	1.46	1.98	1.46	1.72	1.45
19	2.82	4.09	2.12	2.24	2.09
20	20.74	8.79	7.94	8.22	8.01

5   Results

We evaluated the performance of all the models on each microphone in room 4. Each microphone is located in a different place in the room and in different proximity to the main speaker and the three distractor loudspeakers. Hence, each microphone represents a distinct Far-Field scenario. Table 2 summarizes the results in terms of equal error rate (EER). Figure 3 presents the same results in minDCF($P_{\text{target}} = 0.01$).

As illustrated in Fig. 3, all IRL versions show consistent improvement in minDCF results with respect to the Base Model and the finetuned model on all the scenarios. The finetuned model degrades the performance on the closer microphones and improves the performance on the distant microphones compared to Base Model. On all the distant microphones, it gets worse performance than all the IRL models. The tougher the Far-Field scenario gets, the improvement that IRL models achieve is larger - both for EER and minDCF results. On the closest microphones to the main speaker (microphones 1, 2, 3), the base model gets the best EER results compared to the other models. Since the base model was trained on telephony recordings, it is no surprise that it performs the best on the close-field recordings from the closest microphones (their recordings are the most similar to telephony speech). All IRL versions show robustness to all types of Far-Field scenarios. A significant finding from the results is that the Text-Independent IRL model gets pretty similar results to Text-Dependent IRL since text-independence increases the amount of data that can be used for training.

In Fig. 4, we show the distribution of the EER and minDCF1e-2 results on the various microphones. The base model and the Finetuned model have the same median EER value (equals to 3.0) though the Finetuned model has a lower standard deviation. All the IRL models have got similar distribution of EER results and a similar median EER value (near 2.1) that is lower than the Base Model and Finetuned model. At the minDCF plot, we see that the median value of the finetuned model is slightly higher than the median of the base model. All IRL models are better than the Base Model and the finetuned model and have got similar distribution of minDCF results.

Table 3. Mean EER and mean minDCF1e-2 per group of microphones

Microphone group	Base model	Finetune base model on VOiCES	TD-IRL	TI-IRL	DF-IRL
Mean EER (close)	1.92	2.32	1.60	1.74	1.61
Mean EER (far)	15.42	6.98	6.66	6.76	6.76
Mean minDCF1e-2 (close)	0.19	0.23	0.11	0.14	0.14
Mean minDCF1e-2 (far)	0.64	0.56	0.41	0.44	0.41

Another type of evaluation can be made when grouping the microphones according to their distance from the main speaker, close: {1–4, 7–11, 16, 18}, far: {5, 6, 12–15, 17, 19, 20}. The "far" group contains the most challenging scenarios since its microphones are located the farthest away from the main speaker and also closest to the distractor loudspeakers. Some of the speakers in this group are partially/fully obstructed. In Table 3, we see the mean EER and mean minDCF1e-2 of each group. For the "close" microphones group, the Finetune of the Base model got the worst results (worse than the Base Model itself). For the "far" microphones group, the worst results are of the Base Model. All the IRL versions got similar results on both the "close" and "far" microphone groups and were better than the Base Model and the Finetune model. Although the Finetune model improves the performance on the "far" group, IRL models got better results. In minDCF results, on the "close" group, TD-IRL shows 42% improvement and TI-IRL, DF-IRL show 26% improvement compared to the Base Model, while Finetune model degrades the performance. And in the minDCF results of the "far" group, TD-IRL and DF-IRL show 36% improvement, and TI-IRL shows 31% improvement compared to the Base Model.

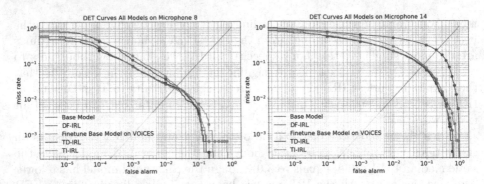

Fig. 5. DET curves of all models on microphones 8 and 14 (loglog scale)

We selected a representative microphone from the "close" and "far" groups and plotted the DET curves for all the models on these microphones (Fig. 5). We visualize the DET curve in a log-log scale since it better visualizes the differences between the models in the small false-alarm regions. The DET curves of all the IRL models are very similar. In Fig. 5 microphone 8, the Finetuned model has a worse curve than the Base model. Figure 5 microphone 14 shows a significant improvement in both the false-alarm and miss-rate scales in all the IRL models. The Finetuned model also shows significant improvement but less than the IRL models.

The DF-IRL model did not significantly improve compared to the other IRL versions, and its training requires "same-text" pairs. Therefore it is not preferable over the TD-IRL model.

6   Conclusions

All IRL versions of x-vector introduced in this paper showed a significant improvement on the Far-Field scenarios and were robust both for close and distant microphones. The Base Model trained on telephony speech with noise augmentations did not perform well on distant microphones. The most significant finding of this paper is that the Text-Independent IRL version achieves similar results to all other Text-Dependent versions. Thus enabling us to use larger amounts of training data of the same speaker from various sources. The (clean, noisy) pairs of utterances are now only required to be from the same speaker and are not required to have the "same-text". We showed that finetuning the Base Model on VOiCES dataset without changing its loss function did not yield consistent improvement. The Finetuned model got worse results than the Base Model on the "close" microphones, and on the distant microphones, all IRL models were better than the Finetuned model.

References

1. Brandschain, L., Graff, D., Walker, K.: Mixer 6 speech LDC2013S03. Hard Drive. Linguistic Data Consortium, Philadelphia (2013)
2. Chung, J.S., Nagrani, A., Zisserman, A.: VoxCeleb2: deep speaker recognition. arXiv preprint arXiv:1806.05622 (2018)
3. Deng, J., Guo, J., Xue, N., Zafeiriou, S.: Arcface: Additive angular margin loss for deep face recognition. In: Proceedings of the IEEE/CVF Conference on Computer Vision and Pattern Recognition, pp. 4690–4699 (2019)
4. Doddington, G.R., Przybocki, M.A., Martin, A.F., Reynolds, D.A.: The NIST speaker recognition evaluation-overview, methodology, systems, results, perspective. Speech Commun. 31(2–3), 225–254 (2000)
5. Godfrey, J., Holliman, E.: Switchboard-1 release 2 LDC97S62. Linguistic Data Consortium (1993)
6. He, K., Zhang, X., Ren, S., Sun, J.: Deep residual learning for image recognition. In: Proceedings of the IEEE Conference on Computer Vision and Pattern Recognition, pp. 770–778 (2016)
7. Huang, J., Bocklet, T.: Intel far-field speaker recognition system for voices challenge 2019. In: Interspeech, pp. 2473–2477 (2019)
8. Ioffe, S.: Probabilistic linear discriminant analysis. In: Leonardis, A., Bischof, H., Pinz, A. (eds.) ECCV 2006. LNCS, vol. 3954, pp. 531–542. Springer, Heidelberg (2006). https://doi.org/10.1007/11744085_41
9. Kataria, S., Nidadavolu, P.S., Villalba, J., Chen, N., Garcia-Perera, P., Dehak, N.: Feature enhancement with deep feature losses for speaker verification. In: ICASSP 2020-2020 IEEE International Conference on Acoustics, Speech and Signal Processing (ICASSP), pp. 7584–7588. IEEE (2020)
10. Kataria, S., Nidadavolu, P.S., Villalba, J., Dehak, N.: Analysis of deep feature loss based enhancement for speaker verification. arXiv preprint arXiv:2002.00139 (2020)
11. Ko, T., Peddinti, V., Povey, D., Seltzer, M.L., Khudanpur, S.: A study on data augmentation of reverberant speech for robust speech recognition. In: 2017 IEEE International Conference on Acoustics, Speech and Signal Processing (ICASSP), pp. 5220–5224. IEEE (2017)
12. Li, Y., Gao, F., Ou, Z., Sun, J.: Angular softmax loss for end-to-end speaker verification. In: 2018 11th International Symposium on Chinese Spoken Language Processing (ISCSLP), pp. 190–194. IEEE (2018)
13. Liang, D., Huang, Z., Lipton, Z.C.: Learning noise-invariant representations for robust speech recognition. In: 2018 IEEE Spoken Language Technology Workshop (SLT), pp. 56–63. IEEE (2018)
14. Mošner, L., Matějka, P., Novotný, O., Černocký, J.H.: Dereverberation and beamforming in far-field speaker recognition. In: 2018 IEEE International Conference on Acoustics, Speech and Signal Processing (ICASSP), pp. 5254–5258. IEEE (2018)
15. Nagrani, A., Chung, J.S., Zisserman, A.: VoxCeleb: a large-scale speaker identification dataset. arXiv preprint arXiv:1706.08612 (2017)
16. Panayotov, V., Chen, G., Povey, D., Khudanpur, S.: Librispeech: an asr corpus based on public domain audio books. In: 2015 IEEE International Conference on Acoustics, Speech and Signal Processing (ICASSP), pp. 5206–5210. IEEE (2015)
17. Richey, C., et al.: Voices obscured in complex environmental settings (voices) corpus. arXiv preprint arXiv:1804.05053 (2018)

18. Snyder, D., Chen, G., Povey, D.: MUSAN: a music, speech, and noise corpus. arXiv preprint arXiv:1510.08484 (2015)
19. Snyder, D., Garcia-Romero, D., Sell, G., Povey, D., Khudanpur, S.: X-vectors: robust DNN embeddings for speaker recognition. In: 2018 IEEE International Conference on Acoustics, Speech and Signal Processing (ICASSP), pp. 5329–5333. IEEE (2018)
20. Villalba, J., et al.: Advances in speaker recognition for telephone and audio-visual data: the JHU-MIT submission for NIST SRE19. In: Proceedings of Odyssey (2020)
21. Yu, Y.Q., Fan, L., Li, W.J.: Ensemble additive margin softmax for speaker verification. In: ICASSP 2019-2019 IEEE International Conference on Acoustics, Speech and Signal Processing (ICASSP), pp. 6046–6050. IEEE (2019)

Author Index

Printed in the United States
by Baker & Taylor Publisher Services

Printed in the United States
by Baker & Taylor Publisher Services